Viking Place Names
of East Lothian

Of Whales and Dwarves

Iain M.M. Johnstone

Tarmagan Press
Edinburgh

Viking Place Names of East Lothian

Published 2005 by Tarmagan Press

3 Piershill Place, Edinburgh, EH8 7EH

©2005, Iain M.M. Johnstone

All rights reserved

No part of this book may be reproduced, stored in a retrieval system, or transmitted by any means, electronic, mechanical, photocopying, recording, or otherwise, without written permission from the author.

A catalogue record for this book is available from the British Library

ISBN 0-9544992-2-0

Printed and bound by Antony Rowe Ltd, Eastbourne

Of Whales and Dwarves

For Rita—and Rona,

with love.

Viking Place Names of East Lothian

Of Whales and Dwarves

Contents

Introduction		9
Brief History		13
Problem Words		21
Chapter 1	*Traprain*	37
Chapter 2	*Hamfar*	50
Chapter 3	*The Garleton Hills*	56
Chapter 4	*Viking Gods*	69
Chapter 5	*Haddington*	77
Chapter 6	*Walk to North Berwick*	84
Chapter 7	*North Berwick/ East Linton*	99
Chapter 8	*Whittinghame / Ninewar*	109
Glossary		117

Appendix	132
Bibliography	147
Biography	149
Index	154

Of Whales and Dwarves

What's in a name? that which we call a rose

By any other name would smell just as sweet.

Aye, maybe so; but it's not your olfactory sense I wish to appeal to, but your sense of who we Scots are as defined by our history, languages and the testimony of our landscape, where the place names can be powerful witnesses—if properly interrogated.

The author has a website with further information on Scots place names and affords the opportunity to comment. The web address is:

www.scotsplacenames.com

Viking Place Names of East Lothian

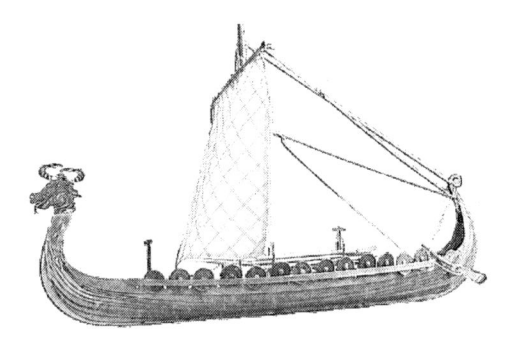

Of Whales and Dwarves

Viking Place Names of East Lothian

Introduction

I wrote a book on Scots place names recently, and what started out as a jolly romp through the more interesting and humorous names very soon became a serious study of the lost place names of Scotland. For over a thousand years these names have suffered in silent anonymity. Some had been brutalised into acceptable modern forms, like Wolfstar, in East Lothian. Sounds romantic, mysterious and even a bit dangerous. Early 13th. century forms (given by the Rev. James Johnston, who thought this name Old English) though showed this name as *Fuylstrother,* which is Norse, *Fúll,* 'foul, stinking' and '*stráðr*', (pronounced *strother*) 'covered in straw', which gives a typical description of an old Scots/Norse farm. Where did the Wolf come from? Well, metathesis is a common linguistic phenomenon, whereby certain letters are transposed. In this instance, *Fúll,* became *Úlf,* 'a wolf' and *stráðr,* became a star.

Now this story is quite interesting, but the significance of it being Norse is that it flies in the face of conventional teaching. If you read books on Scots place names, they invariably say that anything on this side of the Forth that is not demonstrably Gaelic or Brythonic is Anglo-Saxon, since they supposedly made early settlements here. If they did settle, which is highly unlikely, they have left less trace than WMDs in Iraq. Where are the Anglo-Saxon graves? Where the fabulous Anglo-Saxon '*hlaew*' (burial mounds) that academics say are the origin of all the *Law* hills in Scotland? Where the Anglic gods, holy places, business places? NOT ONE! And yet, wherever I looked, I found Norse names. What has been going on? Robert Louis Stevenson called it the 'Anglo-Saxon heresy', (ASH). There are other names.

The purpose of this book is not to go into the differences and linguistic nuances of Old Norse and Anglo-Saxon in any great depth, (away beyond me anyway) but to describe the area within a five or six mile radius of two places near the town of Haddington, with surrounding toponymic (place names) evidence, which could illustrate our Norse heritage with a workable minimum of onomastic, pleonastic and toponymic

Of Whales and Dwarves

dissections. One is the area of Whittingham(e) to the south of the great hill, wrongly named 'Traprain Law', that looks across to the old A1 as it climbs the hill to Pencraig. 'Traprain Law' is over 600 feet high and looks majestic—apart from its eastern quarters that have been quarried by cultural vandals less than a hundred years ago. The other place is called, Ninewar, the name of a large farm, situated to the southwest of 'Traprain Law' and near Whittinghame. Its name used to drive me to distraction for years. Now I smile at both of them.

One final point. Place names of any country are fraught with etymological traps which even the professionals blunder into at times. I think most of them would admit to that. Genuine mistakes are made, and I have no problem with this, having had some experience now with making more than my fair share.

However, the mere suspicion that place names have been suppressed or 'improved upon' for political reasons is something to be less than happy with. It is a hard enough task to undertake research of ancient names and history without factoring in the poisonous effect of 'manipulation', but it has to be addressed. Spin has deservedly been given a bad name, and should play no part in our history, which now has so many voices clamouring for its raised status in our schools. Good luck to them. But take great care with what is being taught—and who is doing the teaching.

Abbreviations used in this book.

Norse-N (Basically Old Icelandic, occasionally Old Norwegian); Sc-Scots; G. Gaelic; B. Brythonic; O.E.- Old English; Anglo-Saxon-A/S; B/T-Bosworth and Toller Anglo-Saxon Dictionary; ASH-Anglo-Saxon Heresy; C/V- Cleasby/Vigfusson Icelandic Dictionary; DOST, Dictionary of the Scots Tongue; SND-Scottish National Dictionary, and C.S.D.- Concise Scottish Dictionary; C.O.D-Concise Oxford Dictionary

Of Whales and Dwarves

A brief historical background

There were almost certainly Scots, that is, Celts, Irish Gaels, who settled in Scotland before the main settlements in Dalriada on the west coast of present day Scotland, in the 5th. and 6th. centuries, apparently with the permission of the Picts. There is also evidence to show that Gaels settled in the area round Dundee at an early stage and have given their name to the city.

The place we now call Scotland, *Alba* in Gaelic, means 'white', which is what the first Celts who landed on these shores would have seen in the cliffs of Dover, and the name was originally applied to the whole island from the Isle of Wight to the far north of Scotland including the Orkneys. The original Celts were Brythons or Britons, arriving in Alba, perhaps about 500 B.C. The evidence shows that their pre-Roman name was Pretanni or something similar and they could have been called Pritons, but became Pridain/Pryden, which certain newcomers, to these shores, Anglo-Saxons, took as meaning, 'Picts'.

To complicate matters we must note that these Brythons, were known as Cruithne to the Scots and Irish. These Brythons had replaced or coexisted with now unknown previous inhabitants of these islands.

The Picts were Celts, a branch of the Brythons, possessing greater linguistic similarities with the Brythons than with the Scots. They arrived later than the Brythons and the Gaels, possibly at the latter end of the B.C. period, and settled in the Hebrides, Orkneys and northern mainland of Scotland. Eventually they moved southwards and took over the hegemony from the Brythons, before succumbing to the Scots and Norse in the 9th century. At this time it is reckoned that Gaelic was the language understood, if not spoken, by all the Picts and Brythons who had merged with them. Norse would also be one of the main languages of Scotland, as the place names show.

The connection between Gaels and Brythons becomes clearer when you understand that one of the major differences between these two Celtic peoples was the use of 'P' and 'C'. The original sound which gave rise to these letters developed as a 'P' by the Brythons, (Pryden) and the 'C' sound pronounced as a 'K' (philologists refer to it as a 'Q' sound) by the Scots. *Prenn* in Brythonic, and *Crann* in Gaelic both meant trees, and the similarity is obvious. Similarly, *pen* in B., and *ceann* in G., both meaning 'head'. Other words like *dun* G. and *din,* B. both meaning fortress, hill, show the closeness of the tongues, and also how in time they could degenerate into 'toun and ton', probably helped by the Norse 'tún', pronounced 'toon'.

Of Whales and Dwarves

The Norse had been around here for some time but we only regard them as invaders from about the late 8th. century. More of them later.

Similarly, the Saxons, Sassenach in Gaelic, had joined with the Scots and the Picts in the attacks on the Brythons as reported by Roman writers in the mid 4th. century. After the Romans left 'Britannia', at the beginning of the 5th. century, Angles and Saxons, (the Scots always referred to them as Saxons) invaded present day England and established fiefdoms in the south, north, and east of England. The native Brythons were pushed to the west and called, Welsh, which comes from the Anglo-Saxon word *Wealas,* which was adapted from a Celtic source, meaning 'a Celt, Briton'. However in England a secondary meaning developed implying, 'foreigner, slave', which is rather petty, totally inaccurate, and unfortunately spewed out on hundreds of websites. The Norse called the Celts *Valir,*(pronounced-Wallir) indeed they called the non-Frankish part of France, Val-land, and Julius Caesar mentions a Celtic tribe the *Volcae*. The name of William Wallace, the great hero of Scotland, also comes from this Celtic source.

The Angles formed an area they called Bernicia in the mid 6th. century, which is claimed reached to the Firth of Forth by the 7th. century. Now just because you throw your towel on a sunlounger, does not mean you own it. Similarly, Angles may indeed have seen the Firth of Forth, but left no settlements there. Din Eidyn

(Edinburgh) was not besieged or captured in 638 by Angles. There was a battle of Glenmorisain, unidentified but probably in the north of Scotland, between Scots under Domnal Breac, according to the Ulster Annals, an ally of the Bernicians, against the Brythons in that year, which the Brythons won, but no mention of Angles. There also was mention of a siege at Etan or Etain, which has not been satisfactorily identified, but that is all. The Anglo Saxon Chronicle makes no mention at all of this year.

Eadwine of Northumberland had his capital Bebbanburch, modern Bamborough, (the local Brythons had named it Din Guayroi centuries before) burned in 623 by Irish/Scots, and then was killed in 633 in Northumberland, and gave his name to nowhere in Scotland. The academic/non academic howls of anguish from certain quarters are indeed that—academic, since Bernicia quite disappeared in the 7th./8th. centuries, subsumed in an area called Northumberland, severely attacked by the Norsemen in the 8th. century and which was occupied by them in the 9th., 10th and 11th. centuries.

The Scots meanwhile had been expanding their territory at the expense of the Brythons and the Picts, although the latter had some notable victories over the Scots, and indeed in the 8th. century, under their king, Angus, took over from the Scots. It was only in the 9th. century when the Picts had been severely weakened by the incursions of the Norsemen, aided by Kenneth MacAlpin, that the Scots under Kenneth MacAlpin finally

Of Whales and Dwarves

imposed some sort of Scots hegemony, in tandem with the Norse, over great parts of Scotland by 839. There were still areas such as the Brythonic Strathclyde, and the Norse Isle of Man, Dumfries and Galloway, much of the Western Isles, Orkney and Shetland, areas round Aberdeen and probably areas down to the Tweed, including Berwick, which were under Norse influence and settlement. Lack of recognition of this fact has caused lots of etymological confusion and misunderstanding of the origin of the Scots non Gaelic tongue, which most certainly had Norse, not Anglo-Saxon, as a major contributor.

In the 10th. century, as in the 9th. century, the Scots were allied to the Norsemen, and King Constantine of Scots joined with Anlaf, Norse king of Dublin and York in attacking Aethelstan the Saxon king at Brunanburh, possibly at the mouth of the Humber, or more probably, at Bromborough (Brumby), near Liverpool, in 937. Tennyson in his translation of the Saxon poem on this conflict mentions the Vikings and the Scots and says they returned to 'Difelin and Iraland', Dublin and Ireland. This did not stop Anlaf becoming king of York and Northumbria in 941, nor the Norseman Amlaib, who had fought against Aethelstan and had been king of York and Northumbria from 939 to 941. Indeed, Brunanburh, seems to have been a dress rehearsal for the real thing. According to the poem, it is said that the Norsemen bragged of having 'had the better in perils of battle'.

Viking Place Names of East Lothian

The Norse took over much of England until 1066, when Norsemen from France, the Normans, many of whose ancestors had invaded England in previous centuries, took over the whole of England, under William the Conqueror. William was ruthless in putting down any resistance to Norman rule and tens of thousands were slain, especially during the 'harrowing of the north', resulting in many English seeking shelter or bondage in the land of the Scots.

Malcolm Canmore, had strong family connections with the Norse, apart from marrying one, Ingiborg, widow of Thorfinn, Viking Earl of Orkney and other parts, who probably was the person we know as MacBeth. The number of Norse place names in the south of Scotland, as well as every other part, will testify to their presence. Dorothy Dunnett's compelling novel, 'King Hereafter', is a meticulously researched historical work, which is highly recommended for anyone wishing to understand the history of the time and make the Scots Norse connection plain. It will also help to explain how the Scots tongue was based on Norse, not Anglo-Saxon as some would say, and so could never be a 'dialect of Old English'.

The later introduction by David 1, king of Scots, of Norman families, added to a heady mixture of competing Celtic and Norse tongues already there—never mind the Saxon speech (with a Hungarian accent) of Malcolm's latest wife, Margaret, and all her friends and

Of Whales and Dwarves

refugee relatives who accompanied her to the Scots court.

This is a very brief outline, and I am aware of many contending opinions as to the origins of the Scots, but due to the main purpose of this book, must leave readers to investigate other claims.

Viking Place Names of East Lothian

Of Whales and Dwarves

Problem Words

There are a great many 'problem' words in Scots place names which cause all sorts of wrong deductions. I can only deal with some of them in this book. In the Glossary I have occasionally gone into some depth with a few like '**Law**' meaning a 'hill'. Because of the false insistence by some that 'law' comes from an O.E. *hlaew*, 'a burial mound', instead of the Norse Lög, 'Law', a hill where the 'Laws' were read out, claims are then made of the English effect on Scots place names from an early period, from Orkney to the Border. They think it then justifies other spurious etymologies, and histories, which in their turn spawn even more errors. It is a situation that must be exposed. Yes there are similarities between Norse and O.E., but why pick an O.E. derivation, instead of often a more suitable Norse origin, especially when there is no history to back it up? That is mischief making—at the least.

Heugh, heuch and some other variations are found in every part of Scotland. It is applied to a cliff face, a ravine, a narrow cleft, coal diggings, a quarry, a place that looks as if it has been hewn out. The Dictionary of the Scots Language (DOSL), which embraces the

Dictionary of the Scots Tongue and the Scottish National Dictionary, says it is from Northern M.E. (beware when you see the letters M.E.) *hogh*, 'a hill', and O.E. *hoh*, 'heel, projecting piece of land'. The dictionaries give many examples of its use.

A 'Heugh', at Ravensheugh Sands, with N. Berwick Law in background

Heugh is Norse, Högg, 1. 'a gap, a ravine or cut like gap in a mountain' (*fjal-högg*); 2. 'a blow or stroke'. We have *högg-eyx* 'a hewing axe'; *högg-járn,* 'hewing iron' and many more examples from Cleasby and Vigfusson's Icelandic Dictionary (C/V). Furthermore, the

Normans introduced a word *hogue*, 'a hill, mound' which gave itself to place names in Normandy and after 1066 to other places, but this Norse word came from *haugr*, a Viking burial mound.

Haugh, hauch, haw, are often confused with the above. Its meaning is 'pasture land; grazing land often next to a river'. A common expression in all Scotland. The DOSL gives **O.E.** *Halh, healh,* 'a corner, a nook' as its origin. Bosworth & Toller, (B/T), in their Anglo-Saxon Dictionary, give 'corner, angle, a secret place'. Yes, you might well wonder what a secret place, nook or corner has to do with alluvial grazing ground—read on and you will appreciate the problem.

Haugh and its variants come from **Norse,** *Hagi haga* 'grazing land, pastures'. C/V gives examples of *fjall- hagar*, 'fell pastures'; *ut-hagi,* 'out pastures'; the name of a farm, Haga-land in the Landnamabok. It is also present in the *haw* of hawthorn, and also in the lovely town of **Hawick**, *Hagavík,* 'pasture land at the confluence of two rivers', which is hardly surprising in this very Norse border town. Yes I know what *they* say. O.E. *haga,* means 'dwelling in a town, enclosure, a place fenced in'. Quite inappropriate here amongst a number of Norse names, such as Slitrig Water, Norse, *slit*, 'breach, slit'—(*slitan* O.E. equivalent) and N. *ryg, hyriggr, rygg,* 'a ridge' which flows down past Berryfell Hill, N. *Berg*, 'rock', *fjall,* 'hill', and Hummelknows,

uniquely Norwegian Norse, *hummell knollr,* 'smooth rounded small hill', are a small selection.

Howe, how and other variants. A hollow or low lying place and common from the Shetlands to all parts of Scotland. Surely they would not try this one on? DOSL states that it is from Middle English *Holl(e)* from O.E. *hol* 'a hole, a hollow'. How the Vikings put up with this sort of thing, I don't know. Bartholomew's Gazetteer of the British Isles, 1900 did not. Under place-name etymology it has *How, haugh*, Norse, 'sepulchral mound' and *Hob, hope, how*, Norse, 'a recess among hills, a shelter'. Heugh, heuch, was apparently unknown to them.

Howe, etc. is Norse *Hol, hola,* 'a hole, hollow' and *Haugr*, 'burial mound'. E.g. Maes Howe in Orkney, along with Sigurd's Howe, Minehowe, etc. The connection between the two words, *haugr* and *hol* is obvious, so that archaeology must supply the information to point to the proper origin. In the examples given it would seem that the burial mound origin holds, although some question the origin of Maes Howe. Habbie's Howe near Carlops seems straightforward, since the area is a low lying depression within woods. Off course if Habbie (Halbert or Hubert) is buried therein it could be from *haugr*! Anyway, it is Norse. Why? Because of the plethora of Norse place names in the area, history of their settlement, and no history of English settlement.

Of Whales and Dwarves

Knowe, know, is a little round hill, top of a hill. DOSL says that M.E. *Knol(l)* from O.E. *cnoll* is the origin of this word. If they did not say that, they would have to say that it came from Old Norse, *Knollr;* Old Swedish, *knol*, Old Danish, *knold,* and then they would have to rewrite their dictionaries. The C.O.D. (Concise Oxford Dictionary) does list Old Norse *Knollr,* 'a hill top' in their definition and etymology of knoll. Why you will ask does the DOSL give these Anglic supposed origins? Their own origins are from Victorian times, when the ASH (Anglo-Saxon Heresy) was propagated throughout the land. If you read the Rev. James Johnston's, Place-Names of Scotland, first published in 1892, you will find it riddled with examples. He had a knowledge of Old Norse, was an intelligent man, but his education forced ASH on him. A recent book, 1995, by David Dorward, (who sadly died just over a year ago) Scotland's Place-Names, continues the heresy. Just one example. He knows that claims are made that the Isle of May, in the Firth of Forth, is from Old Norse *má-ey*, 'sew mew isle', (Rev. J. Johnston) but then says it is difficult to believe that the Norse penetrated so far south. ASH has stifled and falsified Scotland's place names.

Gate in Scots can mean a road. It comes from the Norse, *Gata*, 'a road'. There is a Norse word, *gat,* 'a hole, opening', from which we get the meaning of the gate of a garden and suchlike. *Gata* is the suffix for many Scots words and place names. Canongate, Nungate,

Hardgate, Sidegate, Broadgate, Watergate and **Luggate** are a tiny selection. There is no competition from O.E. The latter name Luggate, means the 'road to the sea' and is fairly common, usually in connection with a burn or river. The Lugg part comes from Norse, *Löggr*, 'the sea', which gives us other words that the C.O.D. and others are perplexed by. Words like, logbook, captain's log, logging on, etc., are terra incognita to the C.O.D. They say it is M.E. orig. unkn. They also give 'a felled tree, log' as one of the meanings. It is Norse *Lág,* 'felled tree, log'. A different word altogether.

Burn, that very Scots word, that certain types say comes from O.E. *burna*, 'a stream'. We have Old Dutch, *burna*, Old Icelandic, *brunnr*, Gaelic *bùrn* (which probably came from the Norse, since the Scots had intermarried with and settled with the Norse at an early stage), which points to an acceptable alternative to a totally unacceptable one.

Water. Yes the stuff you put in whisky—or is it? The O.E. for water is *wæter.* The Old Norse *veittr,* with a similar pronunciation, means 'to carry, convey, act as a conduit'. O.N. *vátr* (pronounced, *water*), means 'wet, wetness'. In other words when in Scotland we see places named like the Water of Leith, the meaning here with a Norse derivation would mean the place, valley, plain, whatever, that the river was in, was leading to, in this instance, the Port of Leith, which is itself Norse in origin, as are the several similarly named places in

Of Whales and Dwarves

Scotland. The Rev. Johnston mentions under Galawater in the Borders, that in accordance with Border usage 'Galawater means the valley through which Gala flows', which fits in nicely with this Norse concept of the conduit or background to the water that was carried. The town of Galashiels is Norse (see next).There is also the Norman word *Vâtre*, 'water', which appeared with the Normans in Scotland about the end of the 11th. century and doubtless had an influence in the pronunciation in Scotland of 'watter' as in the phrase 'If yer patter wis watter, ye'd droon!'

Shiel, shiels, sheel. Common ending throughout Scotland indicating a hut, shed, shelter. DOSL says N.E.D. (New English Dictionary) postulates an Old Northern (English) form *scela*, cognate with Old Norse *skáli*, 'hut, shed'. C/V says *skáli* is also comparable with Scots shieling, with similar meaning. Now out of this information, place name persons repeatedly say 'English *shiel*' when it is part of a Scots name like Galashiels, in order to further the ASH, consciously, or just perhaps, unconsciously. There are *shiel* names in the area of Scotland which is the subject of this book, names like Gamelshiels, Gumishiels, Henshiels. Gamel is a Norse personal name. Gumi is poetic Old Norse for 'a man'. Hen is Old Norse *hœna*, 'a hen'.

There is also the preference of some, *skjól*, 'shelter, cover, protection' as the etymology of *shiel*. It comes from the Norse word for a shield, *skjöldr*, and seems more appropriate. For some reason, Dorward seemed happy with this Norse etymology for *shiel* and mentions Galashiels in this context—which is many miles farther south than the Isle of May in the Firth of Forth, which caused him to baulk at a Norse etymology because it was so far south.

Biel, bield, beel, bel. Common word in Scots for a shelter or protection of some sort. Robert Louis Stevenson wrote in Kidnapped, "O, let me get into the bield of a house - I'll can die there easier." Whether that is the same as *biel* I cannot say for certain, but I think there is a good chance. Place names of this type in our area of study are, Biel Mill, Biel Water, Belton, Belhaven, Biel Hill, Biel Grange, Biel, all within a few miles of each other. The DOSL has come up with "O. North. (?) *bældo*, W.S. (?) *bieldu, byldu*, boldness (from O.E. *bald, beald*), hence extended to what may result from boldness — *i.e.* protection". Eh? Perhaps. But why not look for a Norse etymology as an alternative?

There is Norse, Bald, as in Baldred, N. *baldredi*, 'bold rider', which also probably has nothing to do with *biel*. This name comes from Norse *bæli*, also *bili, býli*, mostly used in compounds, according to C/V, 'den, lair, farm, dwelling'. Covers everything, shelter, protection, is used in compounds, and it is Norwegian Norse, which is just fine in this area.

Of Whales and Dwarves

Ton, toun, toon, tun. Today, all these words mean a town. Originally, it would have signified an enclosure of a house or farm. In Scotland, this word came from the Vikings as *tún*, pronounced 'toon'. The Anglic *tun,* pronounced as spelt, although some say 'toon', later produced 'ton'. It was certainly pronounced 'toon' in the Norse; we certainly have a preponderance of Norse place names in Scotland (non-Celtic), and the common pronunciation still today is 'toon'. No matter the pronunciation, we do know from Ekwall, that *tún*, was quite common in Iceland as a suffix, and that the Vikings in Scotland spoke the same or very similar language.

Shaw is Scots for a wood. DOSL, says from O.E. *sceaga*, 'copse, small wood, thicket'. It also mentions N. *skaga*, 'a projection'. There is Norse *skaga,* v. 'to project, stick out'. There is also *skagi,* N. 'low cape or ness'. These words would produce a word sounding like *shaw*, and are seen in the meaning in Scotland of 'snout or brow of a hill'.

However, Scots *Shaw*, 'a wood', comes from N. *skóg(r)*, 'a wood'. Examples are *skóg-barn*, 'wood bairn'; *skóg-bjarn,* 'wood-bear'; *skóg-land*, 'woodland'.

The *shaw* of crops is from N. *sjá*, 'to see,' with 'appearing' a derivative.

Shaw is also a common Scots name.

It is astounding how this word and others have been used to project an Anglic presence in Scotland's place names, and the social and political consequences arising, based on dodgier evidence than even an MI5, or whatever, agent could produce. And there is an outcry from many quarters for the compulsory teaching of history in our country. But who would do the teaching? There's the rub.

Lea, lie, ly. Fallow ground, ground left lying, meadow. Nothing of substance here, but often there are claims they are Anglic endings. Not much evidence to go on before the 17th. century. However we have Norse, *lyggja*, 'to lie'; *lýja, lý,* 'to beat, hammer, forge'; *hlið*, 'cultivated slope, farm'; *Li, lie*, is a suffix having this latter meaning in Norway today.

Another N. word, is *ljá*, 'cut grass'. There is also Gaelic *lèan,* 'a lea', *laigh, luigh, lie*, Old Gaelic, *lige,* 'bed, to lie'.

I must mention *Lægi*, which with Fair as a prefix gives **Fairlie,** Ayrshire. Norse, *Fagr,* 'fair' and *lægi,* 'anchorage', befitting a town with an extensive shore, pier and shelter. The Rev. J. Johnston, thought it was from Old English, *léah,* 'a meadow, lea', giving a 'fair meadow'—but that was not his fault, and by the way, in case you get the wrong impression; yes, there are grievous faults in Johnston's various editions of his work, but it is still a treasure of original spellings and sources that can be checked and help us to move onwards in

Of Whales and Dwarves

this very tricky field. Without his 'mistakes' I would have been left floundering with many a place. And I am equally certain that I have made mistakes as well.

There are a number of examples of the ending in *lie* or *ly,* being given an erroneous etymology.

Wick. Town, village, confluence of waters, bay, confluence of a river and sea, a creek, inlet. The usage of this expression has expanded from its original Latin, *vicus,* 'street, village', O.E. *wic,* N. *vík.* The original usage of O.E. *wic,* was ceasing to be used by the 7th. century. When the Norse arrived in the 8th. 9th. and 10th. centuries it was used as in N. *Jorvik,* York. Later usage of *wic,* seems to have occurred, probably as an attempt at resurrecting the Anglic language which had been overwhelmed by the Norse rulers of England. The declension endings of O.E. disappeared and about the time of the Norman takeover, we find an '**s**' being used to denote a plural as opposed to the O.E. ending seen in child, children. This is quaintly described by some as an English (**s**) and so denoting an Anglic origin of words in Scotland. If a *wick* is in present day Scotland you can be almost certain it is of Norse origin—no matter how it was spelt in the 12th. century. For example, Hawick, is recorded in the 12th. century as *Hawic.* However as previously shown it is certainly Norse. It has the confluence of waters to make it a genuine *vík,* plus the pastures, *haga*—and a plethora of Norse names. Prof Nicolaison in his book on Scots place names says it is

Viking Place Names of East Lothian

an Anglic 'hedge farm' from O.E. *haga*, 'hedge' and *wic* 'farm'. A seductive conclusion from the flimsiest of evidence. A hedge farm, and nothing else. What is the significance of the *wic?* Very little. By the 12th. century, 'c' and 'k' were commonly interchanged but sounded as a '*k*'. Could it have been the nearby English monks at Melrose who recorded it so? Probably. Their monastery was set up in the first half of the 12th. century, by David, king of Scots, whose mother had been English and he was the earl of Northumbria.

Interestingly, near the monastery in Melrose where these English monks set up home in 1136, is another '*wic'*, Darnick. It is called *Dernewic* in the 12th. c. O.E. *derne*, 'secret, hidden' plus *wic*, 'farm'. Seems conclusive, but there is something fishy here. Prof. Nicolaison says there is no English record of this usage. Of course if it was a secret, no one would have known about it. What is secret about this place? Absolutely nothing. It is perhaps half a mile from the monastery, on open level ground down to the edge of the River Tweed. Could not have been less hidden or secret and looks ideal for building on, which is just what has happened here. The Gaelic, *Darnaig,* ' bit, piece,' also 'a plot of land on which to build a house', which in turn comes from the Brythonic, *darn*, 'a part or piece' and in context meaning 'a building plot'. Not far from this spot, still in Melrose, we have Dingleton (also in East Lothian), Norse, *denga,* 'hammering that sharpens a scythe' and *lýja* (*lý*), 'the act of hammering'.

Of Whales and Dwarves

Prof. Nicolaison compares Darnick with Darncrook in Northumberland, where he says O.E. *derne* has combined with N. *krókr*, 'a nook, bend,' to give presumably a 'secret bend'? The usage of the Brythonic *darn*, (actually borrowed into O.E. as well as Gaelic) 'a bit, piece' would to my mind have produced a more believable etymology.

Other 'wicks' for which O.E. claims are made include, Prestwick, sic. 12th.century. This Ayrshire town was at the heart of the Norse occupation in the 9th. century, before and after the siege of Dumbarton in 870 in the company of their Gall/Gael acquaintances and allies. It has N. names like Shawhill, 'wood hill'; Lady Kirk; Shields, 'dwelling place'; Tarshaw, N. *tarf*, 'a bull', plus 'wood'; and Gaelic names like Bogside; Raith Burn; Cluan, Gaelic, 'meadow'. It is five minutes from Ayr. Norse, *Eyrr*, 'sandbank leading into the sea'. Why shouldn't Prestwick be Norse, *Preost vík* ? The same can be said for Fishwick, N. *Fisk vík*; Fenwick, N. *Fen* 'bog' *vík,* although the latter could well be a hybrid, with Old Gaelic, *fine*, 'kinship', and N. *vík*, which would have emphasised the name of the district around them, the district of Cunningham, Norse, *Kunningi*, 'kin, friends, acquaintances', plus *heim*, 'home'. The home of the Gall/Gael, (the origin of the name Galloway) well documented in the histories, but seemingly ignored by the Scots place name experts.

Berwick, is Norse, *Beruvík*, 'Bear town'. In the Orkney Sagas there is a *Beruvík*, referring to this town on the Tweed. There is Berwick in Orkney today, with the given etymology, *bera, beru*, 'a female bear'. *They* say, perhaps, but this one on the Tweed is an English 'barley farm'. Quite.

An aside. The other day as I was driving through East Lothian, the radio had a discussion with some persons about releasing wild animals into the empty spaces of the Highlands to give the place some zing. Lynx, bear, wolves, that sort of thing. Several outraged citizens came on giving their frank opinion of these controversial persons. One of them, on being taken to task for suggesting the release of foreign beasties like bears, came up with this nugget of information. Did you know, he nasally intoned, that the Romans used to ship the native bears out of Scotland, via the port of Berwick, which means 'port of the bears'? That shut them up— for a few seconds. Seriously though, there have been many archaelogical digs and studies carried out in the Berwick area, but not a trace of any Anglo-Saxon settlement has been discovered.

Of Whales and Dwarves

There was a time in the 7th. century, when peace and cooperation and no nasty 'nationalism' was the order of the day. Indeed a Pictish king, Talorcan had an Anglic father, Eanfrith, and was brought up at his court, which undoubtedly led to some sort of claims over Pictish land. Whatever the reason, Anglic expansionist ambitions exerted themselves and their king, Ecgbert was killed along with all his army in 685 at Dunechtan after facing the Picts. The victors actually buried Ecgbert in Iona, since he was 'family' of a kind. That is fact. Bede who lived through this period, records that the Picts regained all the lands they had lost (he gives no indication of when they were lost or the extent of the lands) to the Scots and Angles, who had been allies at this time, and this was the case at his death in 735. Of course it took a while to regroup before the Angles were at it again. This time in 756, their Anglic army was wiped out, the king Eadbert, escaped, and became a monk. His son took over and was murdered within six months by his family and chaos descended on those Angles remaining till the Norse arrived in 787 and changed the order forever. This battle and consequences are recorded by Simeon of Durham, but not mentioned in the Anglo-Saxon Chronicle.

Hedderwick. Farm and hill just outside Dunbar, Norse, *Heiðrvík,* 'bay moor'. That is enough—or as they say in Muirkirk, 'I've had a *wheen*(not O.E.) o' that.' Quite Devine.

Viking Place Names of East Lothian

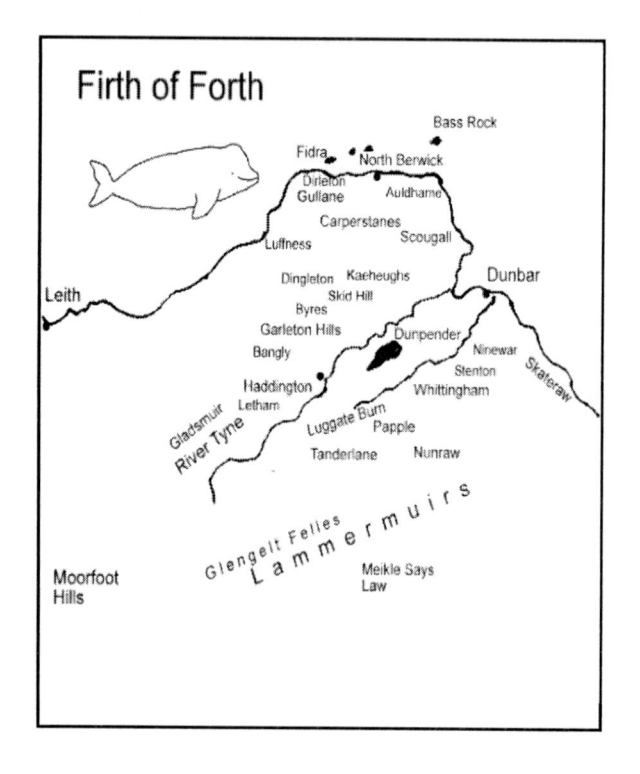

Firth of Forth

[Map showing locations including: Bass Rock, Fidra, North Berwick, Dirleton, Gullane, Auldhame, Carperstanes, Luffness, Scougall, Leith, Dingleton, Kaeheughs, Skid Hill, Byres, Garleton Hills, Dunpender, Dunbar, Bangly, Ninewar, Stenton, Haddington, Whittingham, Skateraw, Gladsmuir, Letham, Luggate Burn, Papple, River Tyne, Tanderlane, Nunraw, Glengelt Felles, Lammermuirs, Moorfoot Hills, Meikle Says Law]

Every name here is Norse, save for Dunpender, Dunbar, and possibly Tyne—and there are dozens and dozens more.

Of Whales and Dwarves

Chapter 1

Traprain

Whittingham(e) is a hamlet situated in East Lothian, a few miles from Haddington, the county town, and facing the great bulk of the misnamed Traprain Law. This is our starting point for a stroll in the surrounding countryside to have at look at the place names and see what clues they give us as to the people, where they came from, their language and activities.

In the 18th. century, Adair's map shows Traprain as Dupenderlaw, while Timothy Pont's map of the 16th. century shows it as Dunpendyrlaw. It is only in the late 18th. century that some cartographical vandal has inserted the name of the nearby hamlet of Traprain, early Brythonic for a 'tree steading', instead of the ancient name of Dunpender, and previously probably, Gaelic, Dunpelder, from Brythonic, Dinpaladyr, which according to Prof. Watson, means, 'Fortress of Spearshafts'. This would have been at least one of the main fortresses of the Brythonic Votadini tribe, who encountered the Romans in the first and second centuries A.D. and made peace presumably with them at that time, because of the evidence of Roman treasure discovered here in the 1920's and residing and on display in the Scottish National Museum in Edinburgh.

Prof. Watson believes this is the place where the mother of the Brythonic (Old Welsh) St. Kentigern (Mungo), Thenaw, was cast down from for being unmarried and pregnant with a child that one day would be the patron saint of Glasgow. Thenaw survived and was cast adrift into the Firth of Forth near Aberlady. This story is taken from Jocelin, a 12th. century monk of Furness, Lancashire. He had been commissioned to look into the lives of the older saints. He actually says that Kepduff was the place she was thrown from, but Prof. Watson thought it wasn't big enough and suggested Dunpender. Today, Kepduff is called Kilduff, since there was a fashion for such name changes some years ago. Makes it look like a saint's abode, 'Church of Duff' or the like, whereas it actually means in Gaelic, 'Black block'. I think it big enough to give you more than a sore head if you were cast from one of its slopes down to the ancient fortress, still visible, at the foot of it.

The Brythons, Old Welsh speaking Celts, seem to have left en masse for other parts in the direction of Wales, in various waves from the 6th. to the 9th. centuries, including Kentigern. After the Siege of Dumbarton by Norse and probably Scots in 870 a.d., there was a major exodus of Brythons from Strathclyde to Wales. It may also have been at this time that many of their fellow

Of Whales and Dwarves

Brythons in the old Lothian, *Loðene,* a Norse name, decided to join them. Some of course would have stayed, like possibly Cospatrick of Dunbar. The family of William Wallace, a Brythonic name, possibly returned to Strathclyde after a period of exile in Wales.

Now the question is who took their place in Strathclyde and the Lothians? The history of this time is murky to say the least, but there are significant pointers that more than suggest that it was the Scots and Norsemen, their allies at the Battle of Brunanburh in 937 and related to the Scots kings since king Kenneth MacAlpin was the father-in-law of Olaf the White, Norse king of Dublin about the middle of the 9th. century. Kenneth MacAlpin had already burned Dunbar five times in the middle of the 9th. century, probably in the company of the Norse.

There is no question of there being Saxons or Angles in Scotland at this time, unless they were refugees from their own people or the Vikings. Saxons lived in places where their name still survives, like Wessex, Sussex, Middlesex. The Angles stretched from East Anglia to Newcastle and had already lost much of their power and influence, especially in the North. Simeon of Durham reports that the English of Northumbria lost most of their army and their king, Eadbert, abdicated in 757 (became a monk) after a disastrous defeat (almost certainly by the Picts under Angus) in southern Scotland, which led to anarchy and confusion for the next hundred years in whatever remained of Bernicia/Northumberland. Nonsense stories of Edinburgh being named after an Anglic king of Bernicia called Edwin, are just that. There was no Edwin. There was an Eadwine, who died in 633, chopped up by a king of Mercia, without giving his name to anywhere in Scotland, because it is etymologically impossible, historically impossible and logically impossible—if you read the history.

It was the Norsemen who restored their version of order with successive kings of York and Northumbria, and established the Danelaw, (see Glossary on Scots word *Law*) which covered most of England, and showed who ruled the land. The very name England, could be Norse (C/V say it comes from Öngull land, Öngull being a Norse personal

Of Whales and Dwarves

name and a name for the Angles. It means 'an angle, fish hook, bent'). After all, Scotland, Iraland, Iceland, Greenland, Shetland, Vinland etc. were named by the Norse. If it had been English, it would have been something like *Anglecyn,* which was a term used by Bede, or *Angle* something. The Saxons of course were not Angles, or indeed Engles (A/S *engle* means Angel), and being the greater number and of the greater influence, would have probably preferred another name. The Scots called England, *Sassan,* land of the Saxons and thus Sassanach for an Englishman. The Irish have *Sasana* and the Brythons, *Lloegr,* 'lost lands'.

Now, I am perfectly aware of the eyebrows ascending heavenwards and lower lips navel gazing at this rough sketch of those times and its being at some variance with the accepted official story.

This is where toponymy can play a significant role. Prof. Ó'Corráinn, of U.C.C. says that "toponymy (the study of place names) is a surly, inarticulate and ambiguous witness, even in the hands of the best counsel", but when the numbers and spread of the inarticulate witnesses are great enough, I think even sceptics must concede some worth to them, especially when there is historical and topographical evidence in support. The area which is under study in this book is a good example.

Norse names abound all over southern, central, northern Scotland and the islands. They have been suppressed through ignorance and/or perhaps for perceived political advantage from succeeding dynasties. The ignorance I can deal with; the political spinning mentality I am afraid is dead wood, beyond help. The ignorance I refer to is the result of an education like mine, and millions more, which was based on 'accepted wisdoms' which no history teacher or officially sanctioned publication would dare to question. It has been a self perpetuating delusion with no one questioning the emperor's lack of apparel—and unfortunately I have unwittingly played my part in this delusion. I must mention several people who have helped me to see the emperor—not a pretty sight—starkers (from Norse *sterk*, A/Saxon is *stearc*).

Robert Louis Stevenson in his last letter to his brother referred to the Anglo Saxon heresy of the official histories, which said Anglo-Saxons had settled here and left their language. It set me on a path which led to Dorothy Dunnett and her great work, 'King Hereafter', wherein she showed, using painstakingly assembled genealogies, how the Earl Thorfinn, Viking ruler of the Orkneys and ten other earldoms in Scotland was better known in his day as the Scots king MacBeth. I laughed at first, but soon came to realise, as did the Scots Historiographer Royal of the time, that it just had to be true. More than 90% of Scotland was under the sway of Thorfinn, and his mother was Bethoc, making him MacBeth, i.e. son of Bethoc. We had a Norse king. The more Norse names

Of Whales and Dwarves

I uncovered, the more relaxed I became with my preposterous findings—everywhere in Scotland!

* * *

One of the places in this area that I visited, Stenton, was fairly well known to me. There is a nearby loch, Pressmennan, in a wooded setting, which was a favourite place of mine for teaching my dogs how to swim. The higher slopes of the wood provide magnificent views to Stenton, North Berwick and the Forth. What I hadn't realised before was the Norse origins of this place.

Old spellings of the 12th. century, give Steinton. This is from Norse, *stein,* (pronounced in the distinctive Scots fashion, *steen* or *stain*) 'stone' and *tún,* 'an enclosure'.

A star pupil at Pressmennan Loch, nr. Stenton

Later of course, the *tún,* pronounced toun/toon, would refer to larger living places, sometimes finishing up as a '*toon*, *toun* or *ton*. This *tún,* was common in Iceland (Ekwall, 1924), where many of our Norse visitors came from via Norway and also Ireland. The A/Saxon word for stone is, *stan*, as in Laurel.

Now if this place was isolated, surrounded by Gaelic or Brythonic place names, then there would be grave doubts as to its provenance, never mind the 12th. century Norse spelling, which could be put down to other causes. There are Gaelic and Brythonic places, like Ballencrief, Achingall, Tranent, Trabroun, Traprain, Dunbar, etc., but these names merge into the local Norse names and in no way overwhelm them. On the contrary, I have found that the Norse names are predominant. And it is in this context that Stenton can confidently be sourced as Norse. Of course you will demand quite rightly for solid proof of these other places.

In Timothy Pont's 16th. c. map can be seen a place in Pressmennan Wood named *Fattlipps*, which does sound humorous and many place name commentators seem happy with that. However, *Fatt* is Norse for 'upturned or bent backwards' and *lipps,* may come from Old Scots *lippie,* 'flax or corn seed measure'. This would then give us a meaning *perhaps* of upturned flax flowers or corn heads.

There is also on this map a Fatlipps (sic) in Midlothian, between Penicuik and Dalhousie near a place Halles, which is probably the same as our Old Norse, Hailes in East Lothian (see page 47). Karkettill, another famous Norse name is a few miles north.

Stenton with N. Berwick and Firth of Forth in background

Just outside Stenton, we find Meiklerig and Meiklerig Wood. This is Norse, *Mikill, mikil,* 'great, tall, large size', *hryggr,* 'ridge', *viðr,* 'wood'. 'V' in Norse is pronounced 'W', and the end 'R', indicating the nominative case, usually disappears. '*K*' is a Norse feature, not found naturally in Anglo-Saxon, which used

Viking Place Names of East Lothian

the Latin *c*. This word *mikil,* is found all over Scotland in various guises as mykel, mukel, mykyll and many more. Old English didn't have a 'K' in their alphabet. Bit of a giveaway.

Close by Stenton is a little place called Ginglet. Strange sounding name and I have no old forms of it. However here goes. There is Norse, *Göngu-líð*, 'footmen, also, help or assistance'.

A short distance away is Spott Wood, Farm, Mill, Burn, etc. This is Norse, *Spotti,* 'bit, small piece'. Anglo-Saxon word is *splott.* Norse for a 'mill' is *mylna*, found all over Scotland spelt similarly and a common Scots name, Mylne. 'Burn' comes (with metathesis) from Norse *Brunnr*, 'a spring, running water'. The Gaelic is *bùrn*, probably from the Norse. Farm is French. Overlooking Spott is Brunt Hill and The Brunt. *Brunt* is Norse 'burnt, barren heath', and incidentally possibly the meaning of Burntisland in Fife. A bit above Pressmennan (Brythonic 'wood of the hill') is Rammer Wood, which is Norse, *Ram(m)r,* 'strong, mighty'.

Robert Louis Stevenson had relatives in this area, and he used to go to North Berwick amongst other places for holidays. Not far from Whittinghame is Stevenson House and Stevenson Mains. On Pont's late 16th. c. map it is *Steenstoun*. R.L.S. had tried for some years to trace his family roots in order to satisfy certain worries but died before succeeding. I think his name is probably Norse, *Stefanson* a fairly common name still today. I

Of Whales and Dwarves

also noticed a *Stefansdottir*. A medial *f* in Norse commonly changed later to a *v*. Stefan is recorded as an Old Norse name, so R.L.S. has no fears.

Within a short distance from here we have Coldale, and several Colstouns. The Norse for coal, and dark, was *kol*, but it is recorded that there is no coal around here, in the Statistical Account of the 18th. century. There is however, the Norse personal name *Kol*, and still today in Scandinavia. What about the *C*? In the 12th. c. Icelandic grammarians laid down rules when C could be used instead of K. These rules seem to have been very flexibly observed, and the result was the names we see today.

On the other side of Stevenson House we have Hailes (Hales in 16th.c.) Castle, situated near Nether Hailes. Nether is *Neðar*, in Norse, and means 'lower', as opposed to *ofarr*, 'over, higher up' as in Over Hailes which is opposite Hailes Castle further up the slope, which is *hallr*, or *háls*, 'ridge, hill', and very common in Scotland. The owner at one time of the castle was Francis, Earl of Bothwell, a Norse name of difficult etymology. However, I suggest, N. *Boð*, 'an order, command, summons, a battle', and *vel,* 'well, good, fine'. There is also *veldi*, 'power'. There are several other possibilities.

He was the third husband of Mary, queen of Scots, and died in a prison in Denmark after being chained to his dungeon wall for some years. His mummified body

Viking Place Names of East Lothian

which lay in a Danish church on display for many centuries was recently interred.

Hailes Castle

Along from Hailes there is Garlabanck as it was in the 16th.c. map of Pont. This is N. *Geir-laukr*, 'garlic', plus *bakki,* 'bank, slope leading to a river, usually'. Today there is Gourlay Bank, in nearby Haddington. The Norse were keen on their condiments. Double 'K' in Norse commonly became *nk.* Whenever you see the word *bank,* in this sense, in early times, you may count on a Norse origin, because there was no Anglo-Saxon word like it.

A bit north of Hailes there are several places with Markle, (15th.c. Walter Bower, Marcle, which he claimed

Of Whales and Dwarves

meant *miracle*), possibly N. *Mark*, 'wood' and *hlið*, 'cultivated hill, slope or farm'. A church here dedicated to St. Mary had been destroyed some years ago.

Heading down the hill from Pencraig we find the village of East Linton. This is B. *llin*, G. *lin*, or Norse, *lin*, meaning 'flax', plus *tún*, or it could be B. *llyn*, G. *linn*, 'pool'. On its outskirts we have Knowes, Norse, *Knollr*, 'rounded small hill, mound' and Hedderwick, farther on towards Dunbar, N. *Heðarvík*, 'heather or moorland bay'.

Chapter 2

Hamfar

The ability to travel in the shape of an animal, *Hamfar*, and become human again was the belief of the Old Norse. They also believed a person could take on the characteristic of the animal. A bear was a favourite subject of this 'shape changing', and is mirrored in the amount of Norse names that have some form of 'bear' in them. Names like, Bjorn, Birna, Biorn, Bersi, Bassi, Ber, either at the beginning or end of the name, all meaning 'bear'.

To a Viking the main attraction of a bear of course is its strength, speed and power; attributes which any warrior wished to have. Some took it to extreme and became known as *berserkers,* literally, 'inside a bears shirt'. Drink, drugs and auto suggestion undoubtedly played a part, and some berserkers appeared to be able to bring on this bearlike state by incantations and chants. Many are described as foaming at the mouth, faces bloated, eyes staring in a maniacal

Of Whales and Dwarves

fashion and impervious to blows from their opponents. Nothing seemed able to hurt them, and they would savage their shields by biting them, attack rocks, trees and even friends on their own side in the conflict. The phrase 'loose cannons' springs to mind. After the battle or whatever, they would become quite weak and take a day or two for recovery, a time when it was not unknown for enemies to take revenge on them.

There are many tales in the Sagas, telling of Odin, and others, changing into, a serpent, a bird, a fish, or almost any type of animal. They could then use this power to travel great distances and trap their enemies. When a Norseman looked at the landscape, he would see it as a living tableau, where rocks, water, trees, had spirits, animal or human, that could materialise. Other cultures also have some of these features. The immediate one I think of is the culture of the indigenous tribes of North America. A disgusting use of this belief was peddled in the 16th.c. by witch hunters in Europe, and especially in Scotland, by people like James V1 and his fellow conspirators, whereby even if a person had an alibi for being in a different place at the time of the specified charge, this could be taken as their ability to remain in a trance or sleep like condition somewhere, while their spirit could travel anywhere and in any shape to commit the alleged crime. One infamous case was the trial of the so-called North Berwick witches where one of them, John Cunningham, a schoolteacher at Prestonpans, was accused of lying in bed at Tranent in a trance while his spirit in the shape of his body was transported to the kirk at North Berwick for a tryst with Satan. He was examined and tortured in the company of James V1 before enduring his judicial murder at a spot which is now under the tarmac of the esplanade at Edinburgh Castle.

Snorri Sturluson, 1179-1241, in his collection of Viking Sagas called the Heimskringla, mentions in the Ynglinga Saga how Odin could change into a bird, wild beast, dragon or fish and travel where he wished. Other Sagas tell of men fighting in the shape of men, then dogs and then turn into eagles, when one is finally victorious. Witches travelling on whales; a whale being sent to spy

Of Whales and Dwarves

on activity in another land; a porpoise fighting a dolphin; these and other fantastical tales were believed by the Vikings and many other peoples in a superstitious age, which is still with us today, and the subject of popular books.

Some people claim that they have experienced these sensations, like flying like a bird and viewing all beneath them. Some believe that in another life they possessed the form of animals. Now these opinions have been around as long as recorded mankind. The Roman historian Herodotus informs us of the Egyptians belief in immortality of the soul and of its passage over time through experiences in many different forms of life. It has been given the name Metempsychosis, and was present in every part of the world, including Australia, North and South America, Africa, and Asia.

Variants of this belief sometimes have eternal damnation for some souls and possible redemption after various trials in others. The Greek philosophers believed in Metempsychosis. Pythagoras believed that animals had souls and became a vegetarian. Plato thought that ten thousand years might effect a perfection of the soul for ordinary mortals, but that about three thousand years would be sufficient for a philosopher to achieve this. Seems fair.

The Norse Vikings believed in immortality if you lived a proper life. For a warrior that meant dying in battle, being chosen by the Valkyries as a hero and then being welcomed into Valhalla, where they could wine and dine before fighting battles, dying—and waking up the next morning, fit and looking forward to the same programme that day.

Odin had two ravens, Hugin and Munin, Thought and Memory. He used to send them out every day to spy over the earth and tell him what was happening, the first flying news reporters. He also had two wolves, Freki and Geri. One was a glutton, N. *Frekr* and the other fierce like a wolf, N. *Geri*. He also rode about on an eight legged horse called Sleipner, the mother of whom was Loki, the evil god (yes, I know, but read the story), who also did all sorts of mischief including getting Balder, the son of Odin killed with some mistletoe. What on earth did those Vikings drink? Mead actually. As in *Meadowbank*, Edinburgh, and many other places.

Chapter 3

The Garleton Hills

Kaeheughs Fort-centre, in the trees- from Skid Hill

This range of hills in East Lothian, extends from just north of Haddington down to East Linton. They are exquisitely shaped, with numerous craggy faces and seductive heughs. From the highest top at Skid Hill, 610 feet, extensive views are presented northwards to the Forth, Fife and beyond. The south exposes the town of

Of Whales and Dwarves

Haddington, nestling alongside of the River Tyne, which snakes its way via East Linton to the sea at Tynemouth. Farther south the full range of the Lammermuirs and the Moorfoot Hills is seen, great gently rounded, almost flat tops, encompassing the fertile fields, farms and villages of much of East Lothian below.

It is little wonder that this area has attracted settlers for thousands of years, and still does. When the Vikings arrived here in the 9th. and 10th. centuries, or earlier, they immediately left their mark in the names of their settlements, their gods, heroes, dwarves, legends, occupations, and names of the striking natural features. Were they friendly settlers or did they arrive with an onslaught of death and destruction? Or did they arrive as friends or at least acquaintances of the Scots?

The history is not laid out in an unambiguous manner, to be carefully presented as evidence. We must approach this question with much circumspection, and as much care must be taken with the place names. The official stance is that the Vikings only settled in the Northern and Western Isles, Caithness and Sutherland, and perhaps some coastal settlements around the Solway Firth, and of course, the Isle of Man. The non-Gaelic or non-Brythonic names then, according to this scenario, means that they must be Anglo-Saxon, and must point to early Anglic settlements in the $7^{th}/8^{th}$. centuries in whichever area they are found.

Unfortunately for this theory, it leaves behind a trail of places with no, or an inadequate, explanation. What Anglic words could explain these names in this range of hills; Bangly Hill, Skid Hill, Byres Hill, Garleton, Ugston, Kaeheughs, Dingleton, Smeaton? None come near, and yet these names are left as meaningless signposts in order to placate a nonsense theory of incorrect history. There are no books, (until now—this is it!), to explain the meaning of these names.

Perhaps the name Haddington, and many more, is named after a wandering Saxon, a favourite ploy of incomers, and the ill-informed websites? Not a chance. Saxons operated in southern England, Wessex etc., and rarely set foot up here. Athelstan, allegedly named in 'Athelstaneford', at the foot of the Garletons, did exist and was a Saxon, but his name is not recorded in this place name, which is Gaelic or Norse, *Aðalsteinn*, (the Saxon name is *Æðelstan* and means 'noble-stone' in both tongues) a heathen name and still a name today in Iceland. The name Haddington, has two similar possibilities, one Norse and the other Brythonic from Gaelic, both of which may be correct—but at different times. The Norse one we are interested in is, *Höfdingitun,* from *Höfdingi,* 'a chief, ruler'. I have dealt with it in the chapter on Haddington.

Bangly Hill rises between Haddington and Aberlady, the old port of Haddington. Any early community of any size has practical requirements for settlement, apart from

Of Whales and Dwarves

the land. Ploughs and implements for building and arms are of course a priority. The Norsemen were noted smiths, (Norse *Smið*) and have produced excellent examples of their art in surviving iron swords, knives, helmets, buckles, nails, horseshoes, iron bands, ploughs, etc. "...iron-cold iron is master of all.", as Kipling so forcefully put it. These items were hammered out in forges and the noise would have been almost constant and heard for miles around.

Bangly is Norse, *Bang*, n. 'hammering', *banga,* v. 'to knock, hammer'. *Lýja, lý,* is Norse, 'to forge'. So we now have something like 'hammering from forging activities' with *höll,* 'a hill'. Check all your English dictionaries and you will find this word is Norse. What about the iron (N. *jarn*) they used to forge, you might say. I'd say it possibly came from the disused iron quarry a few hundred yards away near Skid Hill, which was in use in Victorian times and has now been roughly filled in. A lovely name, hidden for centuries, because of ASH; almost like Pompeii, when you think of the old volcanic area the hills lie in. Perhaps their source of iron was nearer and on Bangly Hill itself. It is a pity that archaeologists have not properly examined this area for Viking artefacts; but then, how could they when they were taught that Norsemen never settled here? If they had been, it might have saved not a few 'developments' in Haddington and surrounding area.

Next to Bangly Hill heading east a few hundred yards, we come to Byres Hill, which rises above Byres Farm.

Viking Place Names of East Lothian

On the top of the hill is a monument which affords panoramic views and thoughtfully provided information tablets for a 360° treat. Byres comes from the Norse, *Byr,* also *Bær,* (the same word as in Bœr War), which simply means 'a farmstead, farm yard and building'. No, Anglo-Saxon does not have such a word. There is A/S, *Byre*, 'a birch tree; a son, descendant; an event', but most unhelpful. The Scots Dictionaries say that the Scots word 'byre' does not come from Norse, but possibly from Old English, *búr,* 'a bower, cottage, inner room'. The Anglo-Saxon Heresy.

Now I must mention at this point what the 'experts' say when confronted by Norse words such as these. First of all they might say that the name is near the sea and that, as is well known, the Vikings raided the coasts and may have left a trace here or there. The other ploy is to say that these Norse sounding places were really named by Danish Vikings who had originally invaded England, settled, became Anglicised, travelled north and were basically English speakers by the time they got to Scotland. I do not jest. When all these excuses seem to be wanting, there is sometimes the final statement that Norse and Old English are so similar, it is hard to tell the difference. Fine, then let us call the language Norse. After all it was the Norse who gave the name England, to the country they had just conquered.

Close by Byres Hill is Skid Hill, with the remains of its iron quarry. I think it is a beautiful hill, and the views are just what you would expect from the highest hill in

Of Whales and Dwarves

the range. It comes from Norse *Skiði,* which is the easy bit. It is where our word 'ski' comes from, (and also 'skid', which will be news to many dictionaries like the OED) and also means bits of wood, which were then used for travel. It also could mean a 'ship' because of the resemblance of a ski-shoe to the shape of a warship. It was also a personal name and the name of a county in Norway. Now, perhaps the word was used to describe something, perhaps iron ore, being brought from the nearby quarry, resting on a couple of runners like skis. It could equally be for the personal name, which is my choice.

Heading eastwards from Skid Hill we come after a climb to Kaeheughs, an ancient and large fort dating back to before the arrival of the Romans. When the Norse arrived here it was farming they were interested in and they have left this Norse name, *Kýr,* (SND think it from O.E. cú) plural of *Kú,* 'a cow' and pronounced as 'koo', which if you had dared to pronounce it like that in a 'Scots' classroom in my day you would have been belted with a leather strap, *ye ken*-another taboo word. Gaelic and Scots were belted out of bairns for a great number of years, till we have today the grotesque bool-in-the-mooth (N. *muðr*) speech of 'Scots' announcers on TV.

The Scots tongue is being taught in many Primary Schools, but unfortunately with an assumption of an Anglo-Saxon background.

Ancient Hill Fort of Kaeheughs

The 'heugh' is the steep cliff face to the right in the following picture, which would have made an excellent defence in its time. Cattle there today however (and humans) require an electric fence round the whole fort to prevent bids for eternal freedom from those so inclined. Heugh, (see Problem Words) you will remember, is from Norse, *Högg,* 'stroke, gap, breach, cliff', which could not possibly have come from any Old English word, although the 'experts' still claim it does.

Of Whales and Dwarves

The north side of Kaeheughs, showing the steep 'Heugh' and Garleton monument on Byres Hill in the background.

Barney Mains farm is close by with many *Kýr* roaming about. Old Norse have *Barn, Barni*, denoting a child or young man and is known in place names. There is also *Bjarni*, 'a bear' a fairly common name. The nearby castle is called Barnes Castle, which may be an adaptation of the former or a reflection of later Norman settlers, since Barnes was a common name among them.

Barnes Castle

If we walk to Athelstaneford from this place and continue northwards we come to a farm called Needless. Now I have no early names for this place, (17th.c. Needless) so the following is a speculative attempt and an excuse to mention a quaint Norse custom. The Vikings could be rather huffy at times and if they were unpleased with someone (no, that would be the last resort) they would sit brooding over their fire in their longhouse carving a likeness of the source of their ire on the end of a post. Their 'art' was often expressed in what can only be described as a pornographic manner, placing their target in incredible positions. This Viking statement would then be planted in the ground of their irksome neighbour to

Of Whales and Dwarves

show their dissatisfaction at some action. The post was called a '*Níð*'. A *Níð-reising,* was the act of raising such a libellous post. The local community would see and understand the significance of such an event. I can only surmise that a *Níð-lauss,* might be the removing of such a libel, since *lauss,* has the meaning of removing or absolution. Why it should be still be here after a thousand years is beyond me. However, there is a little place just over two miles away called by the curious name, 'Jagg', of which also I have no early spellings, so treat this as another speculation. It could well be the name of a racehorse, like Beeswing in Dumfriesshire, which had me scratching my head until I read in the Rev Johnston's book that this place was named after a 19th. c. horse. I digress. Norse, *Jag,* 'a quarrel or squabble'; also *jaga,* 'to harp on about something' and finally *jaga* 'to hunt'. Only possibilities.

Perhaps it was the local blacksmith who had been angry. Not far from Needless is Dingleton. Norse, *dengja,* 'to beat, hammer, or to sharpen by hammering' plus *lý,* 'to beat, forge'. Besides Dingleton in Melrose there is Dingle in Kerry in Ireland. There is also a mention of a Denglynhusse in Kerry in the 13th.c.

Not far from here we have Smeaton, outside East Linton. In 16th.c. (published in 17th.c.) Pont it was Smyrtoun, which may be from Norse *Smyrja, smyr,* 'to anoint', which could well be associated with religious or regal services. Haddington was the centre of a very important Norse centre, shown by the Norse names in the town and its name itself which shows that it was the headquarters of a very important Norseman. Whether important enough to be anointed is another matter. Smyrton is a fairly common name.

The Norse had begun to convert to Christianity from the late 9th.c. onwards. By the year 1000 all Norsemen were at least nominally Christians—or dead. King Olaf of Norway, who held sway over his subjects in other lands had decreed that all his subjects had to become Christian. There were some who claimed their civil rights were being broached, but this was countered with some story of 'social inclusion' or whatever and that the king would personally take part in offering in human heathen sacrifices, all those who would not become Christians. It worked and Olaf was made a saint later.

It is possible, presuming Pont's Smyrtoun wasn't a slip of the stylus, that Christians made use of the anointments available at this place, which is surrounded with religious centres.

Other Smeaton's in the Lothians are derived from *Smiðr*, 'Smith' plus *tún*. Now in Danish it was *Smith* as opposed to Norwegian *Smiðr*. It could easily have been mistaken in the past for the prefix of *Smyrtoun* in the 16th.c.

Before leaving the Garletons, I should mention where the name comes from. There is no one else but myself to guide you, since it is one of those sleeping names. The Rev. Johnston, gave me the clue, with his early 12th.c.

spelling of *Garmeltun*. Pont's 16th. c. map showed a spelling of *Garneton*. Because of his Victorian schooling of course, Johnston had been taught that all this area had been colonised by Angles in the 7th./8th. centuries, and then by incomers from the south at later dates, thus accounting for the Scots 'English' based tongue. This was the dreaded ASH (Anglo-Saxon Heresy) so called by Robert Louis Stevenson, which unfortunately meant he would look for place names in English sources, one of which was he tells us was Searle's Onomasticon Anglo-Saxonicum, 1897. All this did for him with regards to this place was a tentative suggestion that perhaps it came from a Saxon called *Germær*, resulting in 'village of Garmel'.

Garmel today is a Norwegian personal name. *Garm*, was the guard dog of the Norse underworld which was tended by a goddess, *Hel*, daughter of Loki. Putting the two together gives us *Garmel*. I will temper this bold step by also suggesting the simple, N. *Garm,* 'a dog', *leit,* 'search' plus *tún*, 'enclosure', giving us something like 'hunting dog farm', if it is not just a simple farm named after a person called Garmel, no matter the etymology of the prefix.

Yggdrasil - The World Tree

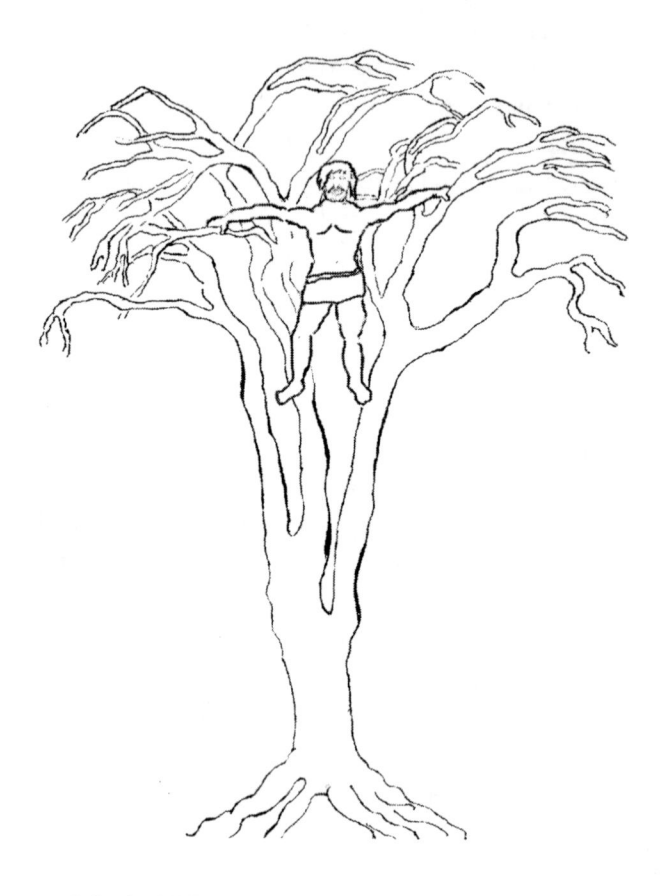

Odin the All-Father, often depicted hanging from the Ash tree, Yggdrasil, imparting his knowledge to the world. This symbolism would have bridged the gap when the Norse embraced Christianity.

Chapter 4

Viking Gods

The Norse religion shared features with other Germanic peoples but had their own distinctive approach. In fact the Norse are the only ones we can really learn from in their Sagas, written down by Snorri Sturluson in the 12th. century, and possibly based on earlier writings as well as the oral Sagas. Roman writers have given their impression of Germanic religion, but not Germans of the time. Odin or Othin is reckoned the Norse version of Woden—or should it be the other way about.

All religions have a creation scenario and according to the Old Norse the world was created by Odin, aided by his brothers Vili, and Ve, after they had destroyed the giant Ymir, the first known creature. They made the heavens and earth out of the body of the dead giant and created their version of Adam and Eve called Ashr and Embla. This whole complex was supported by the world tree, an ash, called Yggdrasil, which stretched from the bowels of the earth up to heaven.

In the roots of this world tree was a well, in a place called Jotunheim, (a fort in Berwickshire, wrongly named at present "Edin's Hall" is named after this place), guarded by a giant Mimir, since it contained all the wisdom of the world. Odin, although he was a god had limitations, and wanting to know as much as possible, visited this place and agreed to part with an eye so that he could drink of this fabulous well. No, I don't know which eye it was and there is some debate about it.

One eye less but with a mind overflowing with knowledge he returned to Valhalla, his palace in Asgard, the home of the gods of whom there were about a dozen. He furthered his quest for knowledge by using the services of two ravens, Hugin and Munin, (Thought and Memory) whom he sent out every morning to look over the world and report to him what was going on. He is sometimes represented hanging on the world tree in which state his wisdom can be imparted to the world. He also had two wolves who were Geri, and Freki (Ravenous and Ferocious) and an eight legged horse called Sleipnir, which is thought to represent four men carrying a coffin.

Odin's wife was Frigga, after whom Friday may have been named. But some say it was named after Freya, the sister of Frey, both of the old Vanir gods who were vanquished by the Aesir, the gods of Odin, which makes it less likely.

Balder, also Baldr, Baldur, (doubtless changed to Baldred after the Norse became Christians) the Beautiful, the son of Odin and Frigga, was the favourite of all gods. He was the god of light, peace, happiness. No wonder he was popular. His mother loved him very much but had a premonition of harm coming to him. She commanded all the living species, whether animal or plant

Of Whales and Dwarves

to swear they would never harm him. The mammoth task was eventually completed. However, the lowly mistletoe plant had been overlooked and had not taken part in the mass vow of compliance with Frigga's demands.

Loki was a low type god. In fact, he was a giant who was the foster brother of Odin and stayed with the gods. An impish clever type, totally untrustworthy, and unpredictable, something like the Fire he was associated with, but no doubt good company for some drunken oafs, of which there was great abundance in those days. Loki discovered the secret of the mistletoe and had a spear like wand made from it. All the gods knew of Balder's protection from all nature and used to sport with him by throwing spears, axes, boulders, etc., which of course just bounced off harmlessly. Balder had a blind brother, Hoð, who was the instrument chosen by Loki to throw the lethal wand he had prepared. Hoð was led to his brother, threw the wand and Balder died at once.

All the gods were devastated. They prepared Balder's ship Ringhorn with a funeral pyre and laid him on it. His wife died of a broken heart and was laid beside him. His horse also joined him. It is recorded that while Odin was conducting these events, a little dwarf called Lit, was running about all excited. Odin tripped him up and kicked him onto the burning pyre, which was swept out to sea on the great ship. Incidentally, it is the only recorded Norse funeral of this type. Normally, Norsemen were buried on land, and at times their ship as well, which is why we have some excellent specimens surviving today.

Frigga's son Hermod, a brave young man, agreed to go to Hel, the goddess of the underworld, and beg her to release Balder because he was so loved by everyone in the world. He journeyed to Hel on his father's steed, Sleipnir, leapt over the ferocious guard dog Garm, and on arriving made his submission. Hel, who was the daughter of Loki, found it hard to believe that everything in the world could have loved Balder, but she said that if he could show that everyone in the world had wept for his death she would release him. Hermod returned and with the help of the gods soon had completed his task apart from one person, an old woman who said she did not love Balder. The old woman was Loki in disguise, and Balder had to remain in Hel forever.

Odin, in the company of all the other gods, set out for revenge and eventually trapped Loki, who had disguised himself as a salmon. Many wished him to be killed at once but the gods decided on a harsher treatment. He was taken to a deep underground cavern where he was bound with iron bands which were unbreakable, and laid on three sharpened stones in a cave full of monstrous insects and creatures, who were nothing but manifestations of the evil thoughts Loki had produced while he was in the world and Asgard. 'Those thoughts that wander through eternity' is a frightful concept and this was to be his punishment. He also had a

Of Whales and Dwarves

venomous serpent dripping red hot poison onto his face making him scream. His faithful wife, Sygn, stayed with him, trying to minimise the poisonous drips on his face. He would remain in this state with no hope of salvation or at least an end to his pain, until Ragnarok, in German, Gotterdammerung, the end of the present time when the gods and giants will fight a great battle and all will be consumed in fire save for a few of the gods and a lucky couple called Lif and Lifthrasr, who will repopulate the world. Thor will be swallowed by the wolf, Fenrir, and the god Tyr, after whom Tuesday is named, will engage Garm, the hound of Hel, and they both will kill each other. Ragnarok (Ragnorök) apparently does not mean 'Twilight of the Gods' but it probably conveys the same message of 'Doom of the Gods'. Sounds quite Wagnerian. It's a pity the Nazis popularised his works, and so tainted them. Still Wagner died in 1920, so it wasn't his fault

The gods' little helpers were the Dwarves. There are famous ones like Brok, the blacksmith of Thor. It is said Thor lost his hammer when he made thunder, and his little workers had to forge new hammers for him. However, it is also said that it was a magic hammer, which always returned to Thor after it had done the business. In the Voluspo, the tale of the prophetess in the Sagas, there is a list of the dwarves. It is amazing proof of the presence here in East Lothian of the Vikings from the Sagas since we have several of their names commemorated in place names from hundreds of years ago. Berwickshire is similarly endowed.

There were other little people, the Alfs, who were overseen by the god Freyr, known as Lord of the Alfs (Elves) in a place called Alfheim. There is a place, Elvingston, in Gladsmuir, which may well come from this source. Freyr had a ship called Skidbladnir and a boar called Gullinbrusti, a golden bristled boar.

Garm- the hound of Hel —or just a farmer's dog

*   *   *

Of Whales and Dwarves

This is basically the background to the social and religious mindset of the Vikings in the 9th. and 10th. centuries, and no doubt for many of them, years after they had become nominal Christians. Now it must also be said that many of them seemed to genuinely embrace Christianity. We hear of a Norseman at the end of the 9th. c. asking the king of Norway's approval to build a church to St. Patrick in Iceland.

There is also Auda Ketilsdotir, the wife of the Viking king Olaf the White, who went to Iceland in the 9th. century and erected a crucifix outside her new home. There are also places like Bunkle Kirk (various names as Bonkle, Boncle in the area) in Berwickshire, which is Norse *Bæna-kall*, 'calling to (God) in prayer'. Falkirk is another Norse name originally for a church and now a town. There are also a number of Viking hogback tomb stones which have been recovered in Christian centres like Abercorn and Govan.

Viking Hogback Tombstones and stone crosses at
Abercorn Kirk in West Lothian

Chapter 5

Haddington

Haddington main town of the county of East Lothian, with an ancient history and great beauty. The name has caused much fanciful speculation. The Statistical Accounts of 1791 and 1845 suggested that Haddington derived its name from Saxon origins but thought the etymology was doubtful.

The Ordinance Gazetteer of 1903, thought that the name came from a wandering Saxon called Hadden, who set up home here; but mentioned a claim that it might be G. *hofdingia-tun*, meaning 'prince's town'. Dorward in his book claims it is 'Hadda's stead'. The Rev. James Johnston shows spellings in 1098, Hadynton, 1150, Hadingtoun, and declares it 'Hading's village', from *Hadda,* one of the founders of the Danish kingdom. I'm not sure why. He does mention that there is another Haddington in Lincolnshire.

I am perfectly willing to accept that perhaps the etymology of 'Hadda' fits in historically with this Lincolnshire place. Why did the East Lothian one have to wait till 1150 before sprouting an 'ing' in its name, plus another 'd' in later centuries? Perhaps, of course, it had indeed sprouted one before. In the 12th. century,

Prof. Michael Lynch, in his 'New History of Scotland' (page 63), tells of three chaps in Glasgow. One was John of Govan, another was Ralph of Haddington, and the other was an Englishman, Robert of Mythyngby. Mythyngby is the modern village of Miningsby, (a Norse name), not many miles from Haddington in Lincolnshire. I trust Ralph and Robert got on well together in Glasgow. Of course I'm not suggesting that Ralph had rushed back to Haddington to tell them how to spell the name of their town—I think someone may have beat him to it.

Hadyn is B. for Aedan, (Aidan) the man who founded Melrose Abbey, and set up the monastery of Lindisfarne in the 7th. century. He was an Irish/Scot who trained at Iona. The king of Northumbria, Oswald, had sheltered in exile at Iona for some seventeen years, under the protection of the Scots, and met Aedan there, and wished for his people to become Christians and literate, since the Angles were illiterate at this time. No books, no histories.

Aedan travelled throughout southern Scotland and northern Northumberland, and some places are named after him—we think. Edington (Haedentun, in 1098) and Ednam (Aednaham, 1105) are a couple provided by the Rev. Johnston. The Brythons in this area would have been sympathetic to a man bearing the name of Aedan, never mind his own personal qualities, since the exploits of Aedan macGabran, who fought with the Brythons against the Angles and who had a Brythonic mother. He had also named his eldest son, Arthur, after the famous

Of Whales and Dwarves

warrior prince or leader of the Brythons who had operated in this area and other areas of Scotland. Since this man was revered in his own lifetime, it would be little surprise that some settlements would be named after him in this Brythonic (Welsh) area.

Now many of the Brythons who had lived in the Lothians for many centuries, left these parts in the late 9th. c. after the area was taken over by Scots and their Norse allies. The same thing on a larger scale had taken place in Strathclyde after the destruction of the Brython's capital, Dunbarton, in 870. In Wales today, Hadyn is a popular first name meaning 'Aidan'. There is a church in Pembrokeshire in Wales, **Llawhaden**, Old Welsh/Brythonic *Llaw* meaning 'dear friend' plus *Haden*, 'Aidan', dedicated to St. Aidan, part of the parish of St. Aidan and St. Mary. The town and castle are also similarly named. Now is this mere coincidence of the names Aidan and St. Mary in Haddington?

There are other places in Scotland named after St. Aedan, but they are mainly Gaelic ones. Originally, G. *M'Aodhan,* this has become '**Modan**'. There are **St. Modans** at Falkirk, **Kilmodan**, Argyll, and in Aberdeenshire. It should be pointed out that there were sixteen saints who used this name.

Whether Aedan had anything to do with setting up an early church in Haddington cannot be determined, at least by myself. But if he did, it could explain the numbers of churches dedicated to St. Mary in this area.

Viking Place Names of East Lothian

Now this seems perfectly acceptable, etymologically, historically—and topography is irrelevant, in this case. However, I think that there is a probability of dual naming of this town. The Norse were definitely here in numbers in the 10th. century if not earlier, and there are so many Norse names in and around Haddington, that the suggestion made of *Hofdingiatun* is almost certainly valid—with the correction that this name is Norse and not Gaelic, although Gall Gaels, Norse allies, were also probably here. It is Norse for 'chief, commander or ruler of the town'. It also shows the Norse *tún*, 'town, enclosure', was an accepted form of ending for a Norse/Scots place. Nothing Anglo-Saxon here.

A chief etc., of course would have a guard, and we have **Herdman Flat** today in Haddington, on a fine commanding position, which is *Hirdmanna Flat,* in Norse, meaning 'the level place occupied by the guard of the king/chief/earl or ruler/commander'. The guard or the king may have liked garlic, and a few yards away from Herdman Flat, we have Gourlaybank, Norse, 'Garlic slope'. Fishing would take place at present day, Poldrate, Norse, *Pollr*, and *dráttr,* plural *drætti*, 'pool of draughts of fish'. And sadly, Briery Bank, perhaps Norse, *Brúr, brýr*, 'a bridge' plus Norse *Banka*, 'river bank', giving 'a river bank leading to a bridge', which is what happens at this place. An archaeologist recently looked at this place and unfortunately found no grounds to prevent another unwanted housing development going

Of Whales and Dwarves

ahead. At the entrance to the town from Gladsmuir, (see below) is an area Roodlands, which is N. *Roða Lands*, 'Holy Cross Land'. All this scene is overlooked by the Norse, Garleton Hills (see under Garleton).

St. Mary's Haddington

* * *

Gladsmuir

An area to the west of Haddington. Today it is mainly a straight stretch of the old A1, with areas to the north and south. It is Norse, *Gleða*, 'a kite, type of hawk', Scots *Gled*, and Norse *mýrr*, 'moor, bog'. The area is still notable for the 'kites' that soar in some numbers. Old spellings of similar named places in Scotland, show 12th.c. Gledhus, and Glademoor, showing that A/S *Gliða*, is not a possibility—never mind the history.

Gladsmuir Kirk and Village

Of Whales and Dwarves

Opposite the church at Gladsmuir, is the quaintly named Butterdean Wood. This is from the Norse personal name *Buthar*, seen also in Butterstone at Dunkeld. In the 16th.c. map of Pont, we find another Butterdean in Berwickshire spelt, *Butterdenn*, showing the confusion caused by many names today called *dean*.

After 1066, and even before, many Norman knights turned up in Scotland, with the suffix, Den, Denn or Dene, signifying their Scandinavian origin of 'Dane'. More confusion. Dane was used to describe Danes, Norse, and Swedish Scandinavians, so care must be taken in assigning a place of origin for these incomers. In Ireland for example, all the Viking 'guests' were called 'Danes' although we know they were mainly Norse, that is Norwegian.

Heading towards Haddington, we come to Trabroun, which is Brythonic, 'steading on the hill' and the place where George Heriot (his name is from the Norse district of Heriot in Borders Region) and founder of Heriot's Hospital in Edinburgh, was born.

Farther along the road we find Letham and Spittal Rigg. Letham is from Norse *Hlaða*, 'at barns', and Norse *Spíttal,* 'a hospice, hospital', plus *Hyryggr*, 'ridge' explains the other. Coincidentally, there are plans to build a new hospital on land here.

Chapter 6

A Walk to North Berwick

If we leave Haddington and the Garletons for the time being and head east from East Linton we come to a place and village today called Tyninghame. You will be told it is the Anglic tongue for 'a settlement on the Tyne'. If you enquire what Birmingham means, you will be told that it is 'the homestead (ham) of the people (inga) of Beorma (or similar). They will give you dozens of examples. Tyninghame, however, they say is the 'settlement on the Tyne'. What about the 'ing' you may ask in vain.

It is claimed that the Annals of Lindisfarne gives an early spelling of, *In Tininghami* (sic) with a date 756. The Annals of Lindisfarne were written in the early 12th. century and the record we have is a Frankish one. The next spelling is Tinningaham, referring to 1050 but written in the 12th.c. There is more, but of little import.

We have no definite spelling of this place before the 11th./12th.c. Duncan 11, king of Scots, granted the land round here to Simeon of Durham's church of Durham. It is about this time that we may hear of a 'ham' or 'hame'. I think Simeon of Durham, who is the only named person to have ever suggested the ridiculous Anglic 'Edwin' as

Of Whales and Dwarves

the origin of Edinburgh, has been at the 'ham' here. Whether he has or not is irrelevant. Tyninghame, comes from the Norse, *inn-gang(a),* 'an entrance' and the Tyne. Whether they regarded this place as their *heim,* 'hame' before this time is of little matter.

I only mention in passing, a Norse god of the sea called 'Tyn', and leave you to speculate.

There is also the possibility of a famous Norse goddess, called, Inga, cropping up here. The Norse Vikings seemed to have a highly developed sense of the mischievous in the application of double entendres.

The entrance to the Tyne here leads a few miles away to the Norse chief town of Haddington. As previously mentioned, we have the Norse Hedderwick, *Heiðar vík*, 'moor, heather bay', on the shores. Nearby at Dunbar, we have the Scart Rock, sticking out in front of the harbour. It is Norse. *Skara, skarað,* 'jutting out'. *Skarað,* can also mean 'a row' as of shields. In a boat where the sides were lined with Viking shields the word used was *skarat,* which would also have the meaning of 'jutting out'. There is also Norse *skári*, 'a young sea mew', which would fit in nicely with the Isle of May in the Forth which is Norse *má-ey*, 'Isle of sea mews'. Scart is also a Scots word for a young sea mew. This could be the youngster gazing wistfully at the home of its parents. A sea mew is of course a seagull.

Next to it is a great big pointy rock, called Meikle Spiker, which is Norse, *Mikil Spíkr*, 'great big pointy rock/splinter/spike (C/V do not list this word meaning 'spike' they have it as 'blubber from a whale'. However, MacBain lists it as coming from Norse *spík* 'a spike'.). On the outskirts of Dunbar we have two places Broxburn and Broxmouth which Prof. Watson reckons comes from Anglo Saxon *bróc,* 'brook' because of 12th. c. spelling of Brock; 11th.c. Broccesmuthe, and 16th.c. spelling of Brooksmyth. First of all, Broxmouth is not at the mouth of anything, and the nearby burn is the Spott Burn, probably named by the Norse a thousand years ago or more, which makes us look with askance at Broxburn. I suggest the Norse *Brúk,* 'a heap, especially of seaweed', plus *mið* 'mark, fishing bank'. Broxmouth is several hundred yards from the sea, and seaweed has been used for food, medicines and for the land.

There is also the conjecture concerning the Norse personal name *Brók,* which literally means 'breeches', plus *smiðr*, 'a smith'. It was also the name of the blacksmith dwarf who did some work for Odin in the Sagas. There is a Brockholes in Berwickshire, which also may be named similarly.

In the great bay of Tyninghame we have some curious names. The Sandy Hirst is one. The first part is Norse, *Sand,* but the second is Gaelic, *hirst,* 'danger' (also the proper name of St. Kilda which is Norse for a 'well'). It is quite simply a large sand bank, which creates danger at certain times of the day and night.

Of Whales and Dwarves

Another name is Heckie's Hole. Here we have the Norse personal name *Heka, Heggi,* which comes from N. *Heggr*, 'bird-cherry tree'. It would have been applied to a place first and then the person named after it—I presume. Norse, *Hol*, is just a 'hole, hollow, depression'. Sounds like his favourite place for fishing, which means I must also mention that there is Norse, *Hákr*, 'a kind of fish', —hake?

At the northern end of the bay we have Whitberry Point, which is Norse *Hvít berg*, 'White rock'. The crashing waves at this place certainly make it look white.

Whitberry with 'Haugr' and Bass Rock skulking on right

It is beside Ravensheugh Sands, which is featured in the Problem Words section. Looking down on the sands at Whitberry is what I believe to be a great burial cairn, a Norse *Haugr*. All this area had a great Norse presence for hundreds of years and there is no better position than this place for such a memorial.

There is also a spot here named St. Baldred's Cradle, a whimsy of some bygone age, or another momento of Baldr? I shall return to Baldred/Baldr later.

The Peffer Sands are next, which is Brythonic, *Pefr*, 'shining, beautiful', named after the Peffer Burn which exits to the sea at this point.

Now we come to Scoughall Rocks and the little settlement of Scoughall. This is spelt Scugg Hall in the 16th. century, with earlier spellings of 11th.c. Scuchale, and 15th.c. Scowgale.

The Rev. Johnston thought it came from N. *Skogr*, 'a wood' and O.E. *healh*, 'nook, corner'. Firstly, Norse, *Skogr*, becomes *Shaw*, in Scots and is seen everywhere in Scotland, including the personal name. Secondly, Scoughall sits out in the open near a headland overlooking the Firth of Forth. It is not a nook or corner in any language, never mind Old English.

Skuggibarrin is a nightclub in Reykjavik, meaning Shady Bar. There is also a place in Iceland, Skuggibjörn. Skuggi is also a nickname, (and was on an Icelandic dating agency website five minutes ago) and the Norse were notorious for using them (nicknames) and I think that is the etymology in this place. However, *Skuggi,*

Of Whales and Dwarves

Norse, 'shadow, spectre, shade' may have been used first to describe the rocks here which are just submerged at times. The name was then transferred to the settlement. *Hallr,* in Norse is, 'hill, slope, big stone. *Hallar* is the genitive of *Höll,* meaning a 'large house, hall'. That would certainly make a shadow. Scougal is a common name in Scotland, and Robert Louis Stevenson has one in Catriona—but perhaps it's in Kidnapped on B.B.C., the recent version of which is utter drivel. The Car Rocks and Great Car, farther up the coast are from G. *Carr*, 'rocks'. The Rodgers are near them. Without old versions I can only point out that *Roðgeir* in Norse is a personal name. Now *geirr,* is N. for a 'spear', which naturally was very symbolic (as well as very practical) for the Vikings. It was commonly believed that a warrior who did not die in battle would not be allowed into Valhalla. A religious rite was practised whereby a warrior dying of natural causes, would cut himself on his breast with a spear and dedicate himself to Odin, thus allowing him (hopefully) to his heaven. We find in C/V many names with 'spear' as a part, like Sig-geirr, þór-geirr, Ás-geirr, Vé-geirr *(the holy spear),* and Geir-hildr, Geir-ríðr, Geir-mundr, Geir-Iaug, **Geir-röðr**. This last name, means 'Spear of the voice', which is a possibility for our **Róðgeir.**

It is also possible that it comes from *Rod*, Norse for a 'fish's skin-from the colouring' plus 'spear'. There are several other possibilities, but I'll pass this one for the time being.

Viking Place Names of East Lothian

Leaving spears behind, we enter the bay called The Gegan. Yes, it is Norse, *Gegn*, 'honest, straightforward; short;' "gentle and steady", is the usage in Scotland according to C/V. It seems to be one of the rock free sections of the coast in these parts, so it may be rather appropriate. There is a common family name of Gegan, for which claims are made of origins ultimately from G. *each*, 'a horse'.

*   *   *

At the north east tip of the The Gegan is a rock called St. Baldred's Boat. It is claimed that he used this stony artifice to sail to the Bass Rock where he had a little chapel. Of course no one believes that nowadays I hope. Actually there is no proof that there ever was a saint called Baldred. The Catholic Encyclopedia (on the Net) says he was (but they do not, neither does anyone else, say who did the saying) an Irishman who was the successor to Kentigern in Strathclyde, and born in 543. A typo in the entry says he was born in 643, unless it was a miracle, since he died they say in 607. Now no Irishman, who obviously would speak Irish Gaelic, could have a name like Baldred. There is Norse, *Baldredi*, 'bold rider'. There is also N. *Baldr*, 'the best', which gave rise to the Norse god Baldr, Balder, Baldur. There is Anglo-Saxon, Baldor, meaning 'bolder, and by extension a ruler', etc. C/V say it comes from a different root from the Norse. However, for an Irishman to be operating in a Brythonic area like Strathclyde with an

Of Whales and Dwarves

English name like Baldred (no matter how rare it was) was impossible, not least because the Angles were not Christians at this time!!

There is one Baldred mentioned in the Anglo Saxon-Chronicle. He was a Mercian king, appointed in Kent and removed a while after in 825. That is the record for 'Baldred'. Not a word of this 'famous' saint.

It is claimed he left many houses for monks and churches in Strathclyde as well as ones at Aldhame, Tyinguham (sic) and Preston Kirk. It then states that he had to give up his work after a short rule for the same reason as his predecessor who went to Wales, namely, the disturbed state of Strathclyde at that time.

Mention is then made of an English monk called Baltherus, who died at Lindisfarne in 756 according to Simeon of Durham (Dunholm), who is turned into the Baldred of the previous century. He was also confused with a Bilfrid of the 7th./8th.c.

The Venerable Bede, the famous churchman and historian, died 735, who has left us extensive records of the time, does not once mention a Baldred, saint or otherwise. The legends of Baldred only originate in writings of the 11th. and 12th. centuries, with Simeon of Durham at their heart. The Liber Sancti Marie of Melrose, written by English monks in the **12th. and 13th. centuries** in the monastery of Melrose, granted to them by David 1, king of Scots, mentions a *Balthere* in the 8th.c and the church of *Baldredi* in the 10th.c. which was burned down at Tiningham. But it does not say where this Tiningham was! There is another Tyne of course in Northumbria. It would be incredible if they didn't have a record of some settlement on it. Hoveden (12th.c. clerk to Henry 11) copied this later. That is all the evidence. I repeat—that is all. A note in the Aberdeen breviary of the 16th.c. regarding this person is only evidence of the Baldred myth being perpetuated.

Simeon's church at Durham had been granted the lands of Tyninghame by the Scots king, Duncan 11, towards the end of the 11th. c. A Norman style chapel was built in the 12th.c. on these lands and called St. Baldred's. In case you are wondering, I am stating that I believe Simeon made up these stories to support an English presence in Scotland. It would also have been done here, with the approval of the new dynasty on the throne, because there already was a name, Balder, the Norse god, and this Baldred the English/Scots/Irish saint, would supplant the heathen god.

Why was this place called St. Baldred's Boat? Well, you remember in the section about Norse deities where I described the death of Balder in his ship Ringhorn? This place could possibly be a commemoration of that Saga event. The Norse for a 'boat' is *Bót*.

*   *   *

Gin Head looms up. Is this the local drunk? Hard to say what it is. But of course there always is a Norse word. In this case, N. *Gin,* 'mouth of an animal or fish'.

Of Whales and Dwarves

The headland (look at it on the map) here does have the shape of a fish's mouth—without the need for Gin.

Just along from our gaping fish, at some rocks, we have Podlie Craig, which is from the Old Scots, *Podlok,* 'a type of cod'. It fairly complements Gin Head. Its origins are obscure but the Oxford Dictionary says of the English *pollack* that there was an earlier Scots, *podlock*. Other dictionaries give some variations of *pollack,* concluding that this latter name and variations came from the Scots *podlok*. Are we all agreed?

Leckmoram Ness is a headland at the outskirts of North Berwick. Prof. Watson suggests the first part might be the Gaelic, *Leac,* 'a flat stone, a slab'. Similar words are in Welsh and Irish. This is then suggested to have become in Scotland, a 'gravestone'. They were called Leckerstanes, Likerstanes, etc.

Leg is Norse, 'a burial place'. It comes from N *liggja,* 'to lie down'. The Gaelic, *leac*, could be purely coincidental, with no borrowing from one language to another. A certain confusion has arisen from the practice of putting slabs of stone on graves.

Looking to Leckmoram

Leckmoram then could be a Norse burial place at a *nes,* headland. The 'moram' section is very difficult since I do not know old forms of this name nor any attempts at its name. It could contain *Mær,* a district of Norway. There is also N. *Máva-grund,* Norse poetical usage for 'land of the sea mews—i.e. the sea'. There is also N. *mar-álmr,* (pronounced Marawlam) 'sea reed,' which gives us 'marram grass', which is quite common along these coasts. Whatever it is, it is a beautiful spot.

The Leithies are a collection of rocks like an island at low tide. Leith, the sea port at Edinburgh, is Norse, *Leið,* 'an assembly place; sea road; passage', and is a fine description of my birthplace. On old maps you will see it described as Leith Roads.

Of Whales and Dwarves

This charming gathering of rocks at North Berwick must have been either a meeting place or a well known spot on the way to the harbour proper at North Berwick.

Of course, there is the possibility they were named by a person who hadn't a clue what *Leith* meant.

There is also a Craigleith island on the other side of North Berwick.

The Leithies with Forth in background

There is an area here known as the Rugged Knowes, which is partly used by a golf course. It is rather fetching walking this part. It is Norse *Rögg* (pronounced Rugg), 'tufted, coarse' plus *knollr*, 'rounded little hillocks'. Originally it meant the tops of mountains.

There is an area to the left here called Heugh, which is Norse, *Högg*, 'a gap, or breach—as a result of 'hewing' out the earth'.

*   *   *

We are now virtually in North Berwick. To look at its etymology we can also include Berwick on Tweed, since they are both connected.

The Orkney Sagas refer to South Berwick as *Beruvik*. There is today, Berwick, *Beruvik* in Orkney. Its meaning derives from Norse, *bera, beru,* 'a female bear'. Other contenders for the Scots Berwick are, B. *beruwi,* 'an outpouring of water', with the Celtic stem, *Ber,* 'a flow of water'. Perhaps we have another case where Norsemen come across a native name and adapt it to their language, and meaning. I think this is perhaps the case here, an original Celtic name, superseded by a Norse one. South Berwick, still has modern O.S. maps showing Norse names such as Sharper's Head, *Skarp eggr,* 'sharp edged' plus *Haufið,* 'head'; Ladies Skerrs, *skers,* 'rocks in the sea'; Bucket Rocks, *búkettil,* 'large kettle', all at the mouth of Berwick.

North Berwick is slightly easier. We already have a bear, a he-bear, *Bassi,* (a Viking name of a *Berserkr* warrior) the Bass Rock, who sits in the water gazing at his female companion. The Norse were an especially romantic and poetical people, when not splitting skulls

Of Whales and Dwarves

etc. The 'Berwick' in this instance is 'Female bear town'. The little Mill Burn which flows into Milsey Bay may have justified the 'wick' at some point. Milsey is a French surname and a locality in the Loire Region which may have nothing to do with the origin of Milsey. However, the Norse, the Normans, colonised the north of France and travelled freely down the Loire basin. The ending *ey,* is N. for an 'island'. It is possible it was a Norman who introduced the name here—or a 19th.c. Victorian returning from his hols in France.

North Berwick Law

There is a recent road in N. Berwick which has been named Nungate Road, perhaps by some foreign *in cognoscenti*. The Nuns in question occupied the ancient Benedictine nunnery of St. Mary's. 'Gate' in Norse and Scots, is N. *gata*, a road.

I think the Norse credentials of this town can take care of any claims that an Old English *bere wic* 'barley

Viking Place Names of East Lothian

farm' is a possibility. North Berwick '**Law**', is a Scots word with a Norse origin (see Law in Glossary). Bass, Law Rocks, Hummel Ridges, Longskelly Pt., Fidra, Carperstanes, Dirleton, Gullane, Garleton, Scoughall, are all Norse names in the vicinity of North Berwick. We also have an area of North Berwick, Carlekemp, which is Norse, *karl kempa,* 'champion of the people'.

In the 12th. and 13th. centuries, Berwick was the largest trading port in Scotland. Berwick was *Berwick* in 1097.

The Bass Rock...is that a bear's head, or is it not?

Of Whales and Dwarves

Chapter 7

North Berwick to East Linton

Leaving North Berwick by the south road before the High School we pass Marly Knowe. Knowe is Norse *knollr*, which is very commonly seen all over Scotland as *knowe,* (see Problem Words) 'a little rounded ridge, hill'. Without early spellings of Marly it is unwise to speculate, but here are some suggestions. *Mark-leið*, is a 'path through a wood' and could possibly be the case here. There is also *Mar-líðendir,* meaning, 'sea-sliders, i.e. witches, spirits'. This is a reference to their supposed ability to travel over sea unaided. It is also a chilling reminder of the murders committed on those persons in the 16th. c. known as the 'North Berwick Witches', although not one of those souls came from North Berwick. It was a result of a plot by James V1, king of Scots, and the local minister, Carmichael of Haddington, which set off the monstrous witch hunts that lasted for over a hundred years. The ruinous remains of St. Andrew's Kirk at the town harbour were supposed to be the place where these victims practised their black art. I have written a novel of this terrible time named 'Acheson's Haven' based on the records of their trials. Enough of this sad tale. Of course, *Mark-leysa,* Norse

Viking Place Names of East Lothian

for 'nonsense', may be a more appropriate comment on these mere suggestions.

To the left of Marly Knowe is Gilsland. *Gil,* is Norse for 'ravine', which I fail to see here. There is the N. personal name *Gils,* which is a shortened form of *Gísli,* 'arrow shaft', and according to C/V, was a popular name from about the year 900 in Iceland, where many of our Viking friends came from or later went to.

Going to the left again after a short distance we see Bonnington, which is a common Norse name (and I stress by Norse I mean Norwegian and Icelandic usage, especially the latter) in Scotland. It comes from the Norse, *Bondi,* 'land owner, farmer, freeman, head of a household'.

Heading due south we come first to Balgone, Gaelic, 'Hound settlement' and then shortly we come to Carperstanes. What on earth could this be? Do we all accept that this is a Norse Viking area now? If so (or if not) we must look for Norse symbols. And what was the symbol that the film industry always gave for the Norsemen? Correct, a raven. That fierce, brooding, black creature of whom Odin had two as companions was the symbol of the Norse. *Korpr,* is N. for a 'raven'. *Stanes* of course is the Scots rendition of the Norse word *stein* they picked up. Anglo-Saxon word was Stan, which is why Stenton was spelt Steinton in the 12th.c. so there would be no confusion.

If we walked sharp right we would come to an area known as Fenton Barns. This area was known as Fenton

Of Whales and Dwarves

in the 13th.c., Fentoun in 17th.c. No 'Barns', which were imported later, perhaps influenced by Barney Mains nearby. *Barni,* was a common Norse place name, based on the word *Barn*, 'a child, bairn'. Norse, *Fen,* 'bog, quagmire. Near East Fenton Farm we have a Whin(g) Bauk, a quarry, which comes from Norse *bauka,* 'to dig' and *Hvin* (pronounced Whin), 'a county in Norway'. All the rocks except quartz, freestone in Berwickshire are called *whins*. In Orkney, a curling stone is called a 'whinnie'. In counties of Scotland like Aberdeen, Stirling, Lanark, Ayr, a whin is a boulder, slab, stone. That is as far as I, or anyone else can go. It is obviously Norse, but the actual source word is elusive, unless that district of Norway called, Hvin, is full of granite type boulders. And of course, the Whinbush, which may just be the coincidence of the omnipresent gorse and the omnipresent 'whins', has to be accounted for as well.

Heading south westwards a bit, there is Waughton and formerly a castle of that name. Probably from N. *Veig,* 'strength'. There is also, *Vág,* 'balance, scales, weight'. Maybe this was the local Norse weights and measures centre.

We keep westwards to Whitekirk, a noticeable landmark with its red church. It used to be white, thus its name. It was also a centre of pilgrimage for many centuries and still is today, with the annual ecumenical walk from St. Mary's Whitekirk to St. Mary's, Haddington.

* * *

St. Mary's Whitekirk-Parish of Hammar

The Ordinance Gazeteer of 1903, states from historical sources that *Hamar* was the name of the adjoining parish of Whitekirk before it changed its name to Whitekirk. It is also stated in the Statistical Account of 1845 by the minister of Whitekirk, the Rev. James Wallace, that the parish of Whitekirk takes in the ancient parishes of 'Tynninghame, Hamer and Aldham' and that the church of 'Hamer' was long called Whitekirk, because of the whiteness of its appearance.

Of Whales and Dwarves

Whitekirk is Norse, *Hvít-Kirkja.* Hamar, is Norse, *Hammar*, 'a hammer, a reference to Thor, the god of thunder, defender of the people, the son of Odin, the defender of the gods etc.' How did a Norse pagan god become identified with a Christian parish? Well in the first place, it must be remembered that the Norse became Christians by the end of the 10th. c. They had no choice. King Olaf of Norway had ordered all his subjects, whether they stayed in Scotland or any other place that dire consequences would be visited on any dissenters.

Of course, some Norse had become Christians since the 9th.c., so it was not as if it was a religion they had little knowledge of. However, no doubt many still hankered after the old religion. Also, Christianity had taken over winter solstice pagan festivals and turned them into Christian ones, like Christmas, Easter etc. In this way, breaking with the past would not be so painful— and they would still get their festivities. Even today, this is a major part of religious festivals for a lot of people. They just got drunk and danced under a different banner.

Baldr would become Baldred and declared a Christian saint. The *Jól Blót*, would become Yuletide. The *Vár Blót,* would become Easter. And Odin, the All-Father, would become the White Krist of the Norsemen. A slight digression for our tipsy inclined friends. A phrase which is heard and understood throughout Scotland (apart from our news readers) is someone 'going to get *blottoed*, and countless variations. This is mainly what happened at these Norse *Blóts*. They drank too much and behaved in an outrageous fashion. Some drank themselves to death. They were the real men. These activities are well recorded in the Sagas. The Concise Oxford Dictionary lists the word but thinks it might come from Icelandic *blettr*, a 'blot'.

Ham(m)ar may be one name that was acceptable until a time came for a change of name. There are other Thor names about which have survived. There is Thurston and Torness outside Dunbar. There is also the Thorter Burn which runs outside the Cistercian monastery of Nunraw, which itself is Norse, *Nunna,* 'a nun' and *rá,* 'corner, nook'. Note it is not a 'row'. And once a week we have Thor's name commemorated on 'Thursday'.

Prestonpans, not far along the coast towards Edinburgh, used to be called Hamer/Hamar and there is still Hammer House there today. It was changed when imported monks from Yorkshire set up Newbattle Abbey of St. Mary in the time of David 1 in the 12th.c. when Hamar became Preston, Norse, *Prestr tún.* Later it became Prestonpans. Today we find a Prestonkirk, also known as Prestonhaugh on the edge of East Linton. Prior to the reformation it was known as Linton and the Haugh, Norse, *Haugr,* 'burial cairn, a mound' or Sc. 'meadowland, often near a river'. Probably from Norse, *hagi,* or hag-jörð 'pasture land'. A claim was made in the 18th.c. Statistical Account for a connection with a 'St. Baldred'. The 1790's document written by the Rev.

Of Whales and Dwarves

Daniel McQueen, stated that in the 'Saxon Annals' this place was referred to as '*Eclesia Sancti Baldridi*'. This 'record' then reports a Saxon *irruption* in the 8th.c. into East Lothian when they burnt the church of *Sancti Baldridi* and the adjoining village of Tyningham about a mile from this place. Remember I earlier pointed out that *Baldridi*, is Norse for 'bold rider'. Rev. McQueen then tells us of places near the present church called St. Baldrid's Well and St. Baldrid's Whill, which latter name he explains as meaning a 'pool or eddy' in the river. Norse, *Hyl* means a 'pool or deep place in a river'.

Now he does not explain what the 'Saxon Annals' are. I guess he means the Anglo-Saxon Chronicles, which were only collated in the 17th. century—but of course they claim they were copied from genuine documents that were destroyed by fire after the collation. However, the Anglo-Saxon Chronicles make NO mention of a St. Baldred. One mention of the aforesaid king Baldred of Kent in 823 being deposed. Not a word of Tyningham, or any variation of it, the nearest being the mention of the Tyne in England in 1066.

The parish Whitekirk/Hammar united with is called Auldhame, now. In the 11th.c. it was, Aldham. It is Norse, *Aldhammar,* a personal name, still today; or possibly *Ald heim,* 'Old Home'.

Now *Hamar,* is Norse for a 'hammer' and is also symbolic of Thor, and there is doubtless a connection here. Thor's hammer, is still worn by many people in this present age.

Viking Place Names of East Lothian

Steading at Auldhame

The other parish in this trinity is Tyninghame. It is from Tyne plus Norse, *Inn-gang(a),* 'an entrance' to the Tyne. To say it is from Anglic *ingaham* is obscure. *Ing,* is an Anglic patronymic, meaning 'of, from' etc. Like Birmingham, 'the village of the sons or people of Beorm', or whatever. If Angles had a settlement here they would have named it something with a person's name. Of course you will find that some try to contort it to mean 'the village of the people on the Tyne', which distorts the patronymic sense. Why didn't the Angles have a

Of Whales and Dwarves

Tyningham(e) on their River Tyne? No planning permission?

The entrance to the River Tyne

To summarise then. There are no known Anglic settlements with an identifiable patronymic in this area in the period before the Vikings arrived at the end of the 8th. century. There are no Anglic or Saxon *hlaew*. No A/S skulls, swords, deities, river names, hill names. No names of any Anglic saint of this period. Not a single trace of Anglic settlements, in the form of ditches, housing, roads etc. There is no history of any reliable sort to show there was any Anglic presence of more than a temporary nature.

However, no matter the known history or the linguistic gymnastics, the Norse etymology is at least as valid as the Anglic, and when history, topography and the abundance of Norse names in all directions in this area are taken into account, it is perverse to advance Anglic claims. Especially when we look at the nearby example of Whittinghame and Ninewar.

Of Whales and Dwarves

Chapter 8

Whittinghame and Ninewar

Whittinghame is said to be one of the three certainties of Anglic 'ingaham' names in Scotland, which are meant to show categoric proof of Anglic settlement at an early period, sometime about the 7th./8th. century. The definition is usually of the form 'the village or home of the sons or family of White'.

One of the proofs that it is Anglic they say is because some locals pronounce it as '*Whittinzhim*'. I will not take part in this local pronunciation evidence. But I will point out that the nearby Athelstaneford, is pronounced by the locals as '*Elshinfird*', which is only testimony to the strength of the local beer. It also does nothing to detract from the fact that Athelstaneford is either Gaelic, or Norse, *Aðalsteinn* plus 'ford.

Another 'proof' is of similar places in the north of England. Proof of what? That there is no written evidence of the eponymous 'White'? O.K. I'll accept that. There is no written evidence of anyone called 'White', so they are similar.

Whittinghame in East Lothian has a great sprawling beautifully shaped hill, shaped liked a beached whale, which commands the district of Whittinghame. Time and again, people comment on its whale shape. East Lothian Council issue Parking Permits in the summer with a drawing of a whale on the front—admittedly frolicking, or whatever they do, in the Forth.

Surprise, surprise. The Whittingham just over the Border near Alnwick has a great sprawling mass of a hill, not beautiful like the East Lothian one, but still a great commanding whale shape overlooking this district. It can be seen for miles. The locals here pronounce the place as Whittinzhim. No, none of them have heard of a family called 'White' round here.

Whittingham, near Alnwick

It is a very pretty village, with actually another hill on the opposite side which could be said to be whale shaped. The other Whittingham is in Cumberland, and

Of Whales and Dwarves

checking it out on the O.S. map, it looks surrounded by hills. I have not visited it, but I wonder if it has a whaleshaped hill looking down on them. Something like the one on this or the previous page.

Alternative Whale near Alnwick

There is also a place, Whiting Bay, in the Isle of Arran, a well known Norse watering hole. Of course this isn't one of the lost sons of White. *Whiting* is a pure Norse word. So is *Geit fjall,* 'Goat Fell', the highest point in Arran.

So what about the recorded evidence of these Anglic 'Whites' in England. Actually they refer to this possible person as *Hwita*. Should be some *Hwitas* about. Not so. It is only <u>presumed</u> that there was a *Hwita*, otherwise how could you possibly explain Whittinghame? This is where the whales come in.

Now what is all this talk of whales about? What could that have to do with a name like Whittinghame? Everything.

Of Whales and Dwarves

Whittinghame, Norse, *Hvítingr Hamr*— Whale Shaped

Viking Place Names of East Lothian

The Luggate Road and Luggate Burn are in front of Dunpelder. The burn leads down to the sea at Dunbar at Belhaven Bay. Luggate, from the Norse, *Lögr*, 'the sea' (which also gives 'Captain's log, log book etc. A surprise no doubt to the O.E.D.) and N. *Gata,* 'a road'. Not that I am suggesting the whale came up that way. This is *hamfar,* on a large scale. It would certainly have made those Vikings think of their travels from home in Norway or Iceland and the sights they had seen.

Now, there may still be some who are not convinced by the abundance of Norse names all over this area, or the Norse belief in *hamfar*, to back up this claim of the district being named by them in this fashion. On record are names such as *Úlfhamr*, 'Wolf shaped'; *valshamr*, 'falcon shaped' in the Eddas; *arnar-ham*, 'eagle shaped'; *faxa-ham*, 'horse shaped'. The *ham*, part has the other meaning of 'skin', from which the meaning of 'shape' derived.

They may be unimpressed by the fact of the evidence of a large 'whale' as opposed to the hypothetical existence of an unrecorded family of Anglic 'Whites'. There are of course many recorded names of Norse 'Whites'. Olaf the White, (*Olafr Hviti*), *Hvit-beinn*, 'White Bone' in the Landnamabok, are but two. I do not think they are involved here, but at least they are recorded. '*Ing*' is also a famous Norse god of peace and fertility and forms many names. They may sniff at the fact there are no Anglic graves, or place names of an early age, whereas I have supplied many showing a

Of Whales and Dwarves

Norse presence. They can point to no historical time recorded even by their Anglo-Saxon Chronicle as to their extensive settlements, which they surely must have had if they claim that the early Scots non-Celtic tongue came from Anglic settlers of an early age, 7/8th. century or later; whereas in the History section I have given dates and places of Norse presence in these parts. I have shown them the Old Norse *Hvitingr*, recorded in Cleasby and Vigfusson with the meaning, 'a type of whale' (I think it's a Beluga). If they are still unimpressed, there is little I can do.

But part of the little, are dwarves. Odin's little helpers, minus poor Lit. They were always around so if they were here looking at the great whale, that might help.

To the right of the photograph of Dunpelder, is a large farm called Ninewar. The one that used to drive me to distraction every time I passed it. Then after many attempts at solving the riddle of its name, one day I was looking at Pont's 16th.c. spelling of this place. It was spelt, *Narwar*. At the same time I knew that there was another Ninewar in Berwickshire, so I looked at this one. It was spelt *Nainnwarr.* So War(r) had remained constant, and yet *Nar* and *Nainn* had both become 'Nine'.

Many false trails were followed after this, till one day while reading Snorri's Edda, Voluspo, the tale of the Wise Woman's Prophecy, there they were. *Nar* and *Nainn*, both together, in a list of dwarves. War(r) then appeared as the Norse goddess *Vár.*

Two Viking age dwarves from the Icelandic Viking Sagas and a goddess, appearing on a 16th. century map, drawn up by Timothy Pont, to whom I record my appreciation of the sterling work he did at the time to preserve some of Scotland's heritage.

After a thousand years or more our little friends have appeared and are pointing their fingers at their giant companion and the history that can now be re-examined. The source of the Scots non-Celtic tongue is not Anglicised Danish, nor did it arise from non-existent Anglic settlements. Well done, little ones—and big ones.

Now is that not a whale of a tale?

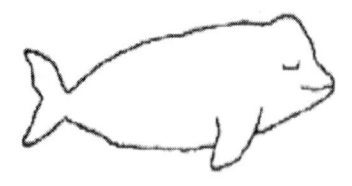

Glossary

Af, N. off, and of. Still commonly heard in Scotland pronounced 'Aff, as in aff his heid.'

Ay, ey, N. an island. e.g. Anglesey, 'Öngull's Isle- Öngull was a Norse personal name and C/V say that this is where the name England comes from'. (O.E. *eà*, 'an island' pron. 'eh oh').

Ayr, eyrr, N. a beach, sand, gravel bank of river or sea

Balder, Baldr, N. son of Norse gods Odin and Freya who is remembered in street names in N. Berwick and a couple of rivers in Yorkshire—I believe he was substituted by the fictitious Baldred in the 11/12th.c.

Bann, N. prohibition, interdict

Bassi, N. personal name of a 'Berserker'-Viking Warrior

Bassi, N. a bear

Barlak, Bygg, N. barley

Ber, N. berry

Bera, beru, N. female bear

Berg, N. rock, boulder, cliff

Berr(i), N. in the open, bare

Bil, N. an open space

Birna, N. a bear, as possibly in Birnam by Dunkeld

Bœr, N.1. Farmhouse 2. Farm.3.Town

Bogi, N. bow shaped

Ból, N. an estate, stead, farm

Bóndi, bœndi, N. peasant, farmer, husbandman, master of household

Borg, N. small dome shaped hill, stronghold, city- Edinaborg (Edinburgh)

Brok, N. personal name, name of Odin's blacksmith dwarf

Bruni, N. fire

Brunnr N. spring, well, water-from which, burn in Gaelic and Scots

Buthar, N. personal name, often confused with 'Butter'

By, bie, byr, bær, bœr, bere, N. town, farm, farm yard, farm building (of course that can mean a 'cowshed' although the Scots Dictionaries say it can't. ASH).

Of Whales and Dwarves

Bygg, N. barley

Byggi, N, inhabitant

Bygging, N. rented land

Býli, N. abode, den, lair, shelter

Byrgi, N. enclosure, fence

Carse, Sc. low lying fertile ground near a stream probably from Celtic/Brythonic, cars, swampy, boggy land near river.

Dalr, N, a dale, a valley. Nithsdale, Annandale, Liddesdale, Clydesdale, Noddsdale, etc.

Dœl, N. a little dale, a recess, a dell in Scots

Dub, N. a puddle (Dub is a very common Scots word from Orkney to Berwick, but our Scots dictionaries think it comes from a Fresian source)

Dýr, N. deer, wild animal

Eik, (pronounced Aik), N. an oak tree, also trees in general. Note Scots 'Aik', meaning oak tree

Elli, N. old

Eng, engi, N. a meadow

Falla, N. to flow, run of water, stream, tide; fala, N.(Swedish), a plain

Far, N. passage, ferry

Fá, N. to paint, draw; fáða, painted

Fær, N. a sheep

Fága, N. to cultivate, till. Also to adorn, embellish, to worship

Fell, N. hill, mountain

Fen, N. bog, quagmire

Fiðri, N. feathers

Fiðrað, N. feathered arrows

Flet, N. raised flooring

Of Whales and Dwarves

Fjörð, N. firth

Floi, N. a mossy moor. Often seen as, flow

Fold, N. an enclosure, a fold, field

Gal, N. gale

Gala, N. singing, chanting, screaming

Gall, G. foreigner, stranger, Also refers to a 'Norseman', the Norse who entered Scotland from the west, Ireland, via the Scots king MacAlpin's lands and settled with the Scots. They were known as ***Gall-Gaels.***

Ganga, N. to walk, to go, to run

Gap, N. an empty space

Garðr, Garth, N. fence, wall, enclosed yard, court, house, dwelling, stronghold, castle

Gat, N. a hole, an opening

Gata, gate, N. a road. It did NOT come from Old English *geat,* 'gate'

Gil, gill, N. a ravine

Greyhundr, N. a bitch

Grimr, N. personal name Graham. Also another name for Odin. Seen in Grimsby

Gunn, N. warrior

Hálfa, N. a region or a part

Hálfr, N. a half

Háls, (pronounced *hawls*), Hawse, N. a neck, a ship's rope, a ridge, a hill, the bow of a ship, the end of a rope. There is the Hawse Inn, at South Queensferry, and a fascinating story in Bower's *Scotichronicon* about a 'hawse'

Harðsteinn (pronounced hard stain) N. whetstone, which could be found in a field for farm implements or a fort for weapons

Haugh, Sc. meadowland, often near a river. Probably from Norse, *hagi,* or hag-jörð 'pasture land'.

Haugr, Haugh, How, N. mound, burial mound, cairn

Heiðr, heðar, N. moor, heath

Heugh, Norse, *högg*, 'stroke, blow'. In this instance applied to the land it means a 'cleft, a narrow gap or breach'.

Of Whales and Dwarves

Höfuð, N. head

Hol, Hola, N. hole, hollow. This became the common 'howe' and other variants in Scots. The ASH (Anglo Saxon Heresy) of course makes other claims.

Hóll, höll, hallr, hyll, N. hill

Holt, N. a wood

Hóp, Hope, N. a small bay or inlet, a sheltered bay

Hólm, N. a little island, usually in a river or bay, AND ALSO a river meadow, riverside land. An Old English *holm,* could only arise if history supported extensive settlements of such a kind in the area. It does not.

Hrafn, N. raven

Hús, (pronounced hoose) N. a house

Inga, famous Norse goddess

Ingialdr, N. personal name giving rise to Ingliston, Inglis etc.

Inngang(a), N. entrance

Kaka, N. cake, compact, giving rise to some place names

Kakka. N. heap, or messy pile like wet hay

Kálfr, N. a calf, an island next to a big one. Check the Firth of Forth

Kambr, N. ridge of hills

Kelda, kilda, N. well, bog

Kenni-speki, N. 'faculty of recognition' from which 'kenspeckle', a common Scots word, meaning, 'easily recognised'. Used by Stevenson in his novels but not by the B.B.C.

Kettil, N. a 'kettle, cauldron sometimes used in religious practice; also used with other words to denote chieftain,

Kilting, N. a skirt—so Jack McConnell was right after all

Kirk, from N. kirkja, church. In Scotland, the kirk usually comes first before the name of the dedication as opposed to the English custom, which is the reverse. This gives rise to the questionable, Kirkcudbright of the 12th.c

Kljúfa, N. to split, cleave

Klofi, klyfja, N. cleft in a hill. Cleugh/cleuch in Scots

Of Whales and Dwarves

Knock, from G. cnoc, a hill, showing Norse influence

Knollr, N. a hump, knoll, in Scots, a knowe. Originally, *knollr* was used for tops of mountains. Now it is seen all over Scotland with the above meaning, including the most Norse of all places, Orkney and Shetland. There is also Old Danish, knold, and Old Swedish, Knol

Kol, N. coal, charcoal, also a personal name meaning black or similar

Kollr, N. top, summit

Kot, N. a cottage, hut

Lamb, N. a lamb, also a N. personal name, Lambi, in Sagas.

Law, N. lög. The 'laws' were originally read out by the Norsemen and business dealt with, on a hill. These lawmen, were called lögmenn, (punch this name into a computer and see the numbers of lögmenn from Scandinavia peddling their legal wares from which the personal name, Lamont, in Scotland. A hill, large rock, or as in Iceland, a cliff, the *Lögberg* in Reykjavik, would be the obvious place in an area for people to assemble. It is from this connection that 'law' meaning a hill arose. When you see the name 'Law Hill', in Dundee, this expresses the knowledge of the origins of the word—

and yet commentators call it a pleonasm, an unconscious use of repetitive words. Norse laws in Scotland? How about, 'By(e)-laws' (Old Norse *býr*, 'town,', COED)? Does anyone seriously think all the towns in Scotland with Bye Laws, have the 'Law' part named after an Anglo-Saxon burial mound called a *hlaew*? The place name experts do.

A Scots legal term for an assault on a PERSON in their house is 'Hame-sucken', Old Norse, *heim-sókn*, Bosworth and Taylor's Anglo-Saxon Dictionary. The O.E. version *hám-sócn* means attacking a person's HOUSE. The bizarre claim that 'Law' came from the Anglic *hláew* (pron. *hlah eh*), is unfounded to my mind and many others. I wonder if the Anglo/Saxons thought that the **Danelaw**, Norse, ***Danalög***, that name which denoted they were subject to Norse rule, came from their Anglic word *hláew,* meaning 'a burial mound'. The etymology, the history, and especially the topography, point to 'law' being Norse. It is mischievous and bizarre to claim otherwise, but unfortunately all too common. This word 'LAW' is the main mantra of the propagators of the Anglo-Saxon Heresy (ASH). It used to be one of the pillars supporting all the claims of Anglic place names in Scotland, especially from the 'Scots' place name societies.

Leið, Leith, N. a local assembly, a sea road, a levy, path

Lin, G. B. N. flax

Mark, N. wood, and also a boundary

Mark Steinn, N. March stone, a boundary stone. Thus riding the Marches.

Marr, N. 1. horse 2. the sea

Mikil, mikill, N. great, large etc.

Mýr, mýrr, N. moor, bog, swamp

Neðarr, N. lower down, nether

Ness, nes, N. a headland

Pap, papi, N. priest, pope

Pap, N. a breast

Pollr, N. pool

Port, N. gate

Prestr, N. priest

Raun, N. trial, test

Rhudd, B. G. ruadh; N. rauð

Ryg, rygg, N. a ridge of land

'S' Norman/Norse ending to form a plural.

Scári, scaur. N. a young sea mew, Sc. scart

Skáli, sheil, N. hut

Skart, N. show, display

Skellr (ir), skellig, N. splash, slap, crash

Sker, skerri, N. rock in the sea

Skugge, N. shadow, also a personal name

Smiðr, N. a smith in metal, wood, etc.

Sunda, N. a sound i.e. a strait between two islands

Spideal, spital, G. hospital, inn, originally from Latin

Spítall, N. as above

Stokkr, stock, N. block of wood, trunk, log, support for boats

Stein, steinn, sten, N. stone

Tandri, N. fire

Thorpe, N. a village

Threip, (*þrap*)N. a quarrel, usually about boundary demarcations

Thwaite, N. forest clearing

Toddi, N. a bit, piece

Toft, N. a farm, field

Tún, N. originally enclosure, then meadow, town

Turn, N. tower, as in Turnberry in Ayrshire

Ull, N. wool

Vað, N. ford

Vaði, N. danger, peril

Vel, N. well (fine)

Vella, N. v. to well up, from which 'well', source of water

Voe, N. a small bay

Völlr, N. field, meadow

Wark, verk, N. a fortified place

Wick, vik, N. bay, confluence of waters, village, a marsh and also a place where salt can be produced

Wrath, N. hvarf, a turning point

Of Whales and Dwarves

A few common words from Old Norse

Acorn, acre, aff, (meaning 'off' in Scots), afar, all, arm, baking, ('bakster' in Scots) balderdash, bann, bark, bat, berserk, bield, bilge, blemish, blot, blur, brother, burn, burgess, bush, cake, call, clap, cleg, dairy, dale, dance, daughter, day, dawn, deaf, dell, dirt, down, dregs, dub (a puddle), edge, egg, elder, end, errand, father, filly, finger, firth, flag, flat, flaw, freckle, furlough, gaggle, gain, gale, gap, gate, gill, girn, girth, glimpse, gravy, green, greyhound, groin, gust, half, hare, haste, hawk, hawse, hide (animal hide) hinge, horse, hold, how,(hill in Scots) husband, hustings, ice, ill, kid, kilt, knowledge, inner, lamb, land, lane, law, ling, mare, maze, mire, mistake, moor, moss, mother, muck, near, neat, (cattle), neighbouring, net, nether, night, nose, North, oaf, outlaw, oak, (aik in Scots), pap, plough, poinding(sale of a person's belongings because of unpaid fine), pound, raft, ransack, rigmarole, rim, row, ruck, rump, rust, sale, shingle, sister, shirt, shiver, skid, skin, skirt, skull, sky, sleet, slop, sloop, smudge, snout, stump, tackle, tatter, thrust, thwart, tike, tit,(little bird), their, them, they, tram, trough, understand, wad, wake, want, weir, welcome, whim, whin, whore, wicket (small gate originally), window, wood.

APPENDIX

This is an extract from a contribution by:

GEORGE TOBIAS FLOM, B.L., A.M.
Sometime Fellow in German, Columbia University
1900

SCANDINAVIAN INFLUENCE
ON
SOUTHERN LOWLAND SCOTCH

In Southern Scotland, Dumfriesshire, Eastern Kircudbright and Western Roxburgh seem to have formed the center of Scandinavian settlements; so, at any rate, the larger number of place-names would indicate. The dialect spoken here is in many respects very similar to that of Northwestern England, D. 31 in Ellis, and the general character of the place-names is the

Of Whales and Dwarves

same. These are, however, far fewer than in Northwestern England. Worsaae (19th.century Danish archaeologist)gives a list of about 30. This list is not exhaustive. From additional sources, rather

incomplete, I have been able to add about 80 more Scandinavian place-names that occur in Southern Scotland, most of them of the same general character as those in Northwestern England. Among them: Applegarth, Cogarth, Auldgirth, Hartsgarth, Dalsgairth, Tundergarth,

Stonegarthside, Helbeck, Thornythwaite, Twathwaite, Robiethwaite,

Murraythwaite, Lockerby, Alby, Denbie, Middlebie, Dunnabie, Wysebie,

Perceby, Newby, Milby, Warmanbie, Sorbie, Canoby, Begbie, Sterby,

Crosby, Bushby, Magby, Pockby, Humbie, Begbie, Dinlaybyre, Maybole,

Carnbo, Gateside, Glenholm, Broomholm, Twynholm, Yetholm, Smailholm,

Langholm, Cogar, Prestwick, Fenwick, Howgate, Bowland, Arbigland,

Berwick, Southwick, Corstorphine, Rowantree, Eggerness, Southerness,

Boness, etc. There are in all about 110 such place-names, with a number of others that may be either English or Scandinavian. **The number of Scandinavian elements in Southern Scotch is, however, very great and indicates larger settlements than can be**

inferred from place-names alone (my emphasis—but his words). In the case of early settlements these will generally represent fairly well the extent of settlement. But where they have taken place comparatively late, or where they have been of a more peaceful nature, the number of new names of places that result from them may not at all indicate their extent. The Scandinavians that settled in Southern Scotland probably at no time exceeded in number the native population. (**I disagree. The Brythons and Picts would have been the 'native' settlers, but especially after 870, when we know that mass migrations took place towards Wales, apart from thousands of Brythons taken to Ireland as slaves, southern Scotland was wide open to settlement by the Norse and their Gael allies. The place names suggest that the Norse were the majority.**) The place-names would then for the most part remain unchanged. The loanwords found in Southern Scotch and the names of places resemble those of Northwestern England. The same Northern race that located in Cumberland and Westmoreland also located in Scotland.

There has been considerable confusion in the use of the terms Norse and Danish. Either has been used to include the other, or, again, in a still wider sense, as synonymous with Scandinavian; as, for instance, when we speak of the Danish kingdoms in Dublin, or Norse elements in Anglo-Saxon. Danish is the language of

Of Whales and Dwarves

Denmark, Norse the language of Norway. When I use the term Old Danish I mean that dialect of Old Scandinavian, or Old Northern, that developed on Danish soil. By Old Norse I mean the old language of Norway. The one is East Scandinavian, the other West Scandinavian. The term Scandinavian, being rather political than linguistic, is not a good one, but it has the advantage of being clear, and I have used it where the better one, Northern, might lead to confusion with Northern Scotch.

We have no such records of Scandinavian settlements in Northwestern England, but that they took place on an extensive scale 300 place- names in Cumberland and Westmoreland prove. In Southern Scotland, there are only about 100 Scandinavian place-names, which would indicate that such settlements here were on a far smaller scale than in Yorkshire, Lincolnshire, or Cumberland—which inference, however, the large number of Scandinavian elements in Early Scotch seems to disprove. I have attempted to ascertain how extensive these elements are in the literature of Scotland. It is possible that the settlements were more numerous than place-names indicate, that they

took place at a later date, for instance, than those in Central England.

We know that as early as 795 Norse vikings began their visits to Ireland; that they settled and occupied the Western Isles about that time; that in 825 the Faroes were first colonized by Norsemen, partly from the Isles. After 870 Iceland was settled by Norsemen from Norway, but in part also from the Western Isles and Ireland. The 'Austmen' in Ireland, especially Dublin, seem frequently to have visited the opposite shore. It seems probable that Northwestern England was settled chiefly by Norsemen from Ireland, Man, and the Isles on the west.

I have left his original spellings. In his work he also examines the works of early Scots writers like, Barbour,(14th.c.) Wyntoun, Dunbar etc. He lists considerable numbers of words that are of Norse (Norwegian) origin. Here are some of his findings.

LOANWORDS.

AGAIT, _adv._ uniformly. R.R. 622. Sco. _ae_, one, + O.N. _gata_
 literally "ae way," one way.

AGAIT, _adv._ astir, on the way. See Wall.

Of Whales and Dwarves

AGROUF, _adv._ on the stomach, grovelling. Ramsay, II, 339. O.N.
á grúfu, id. See _grouf_.

AIRT (êrt), _vb._ urge, incite, force, guide, show. O.N. _erta_,
to taunt, to tease, _erting_, teasing. Norse _erta_, _örta_,
id. Sw. dial. _erta_, to incite some one to do a thing. Sw.
reta shows metathesis. M.E. _ertin_, to provoke.

ALLGAT, _adv._ always, by all means. Bruce, XII, 36; L.L. 1996. O.N.
allu gatu. O. Ic. _öllu gëtu_. See Kluge, P.G.(2)I, 938.

ALGAIT, ALGATIS, _adv._ wholly. Douglas, II, 15, 32; II, 129, 31.
See Kluge, P.G.(2)I, 938.

ALTHING, as a _sb._ everything. Gau, 8, 30, corresponding to Dan.
alting. "Over al thing," Dan. _over alting_. Not to be taken
as a regular Sco. word, however. Gau has a number of other

expressions which correspond closely to those of the Dan.

original of Kristjern Pedersen, of which Gau's work is a

translation.

ANGER, _sb._ grief, misery. Bruce, I, 235. Sco. Pro. 29. O.N.

angr, grief, sorrow. See Bradley's Stratmann, and Kluge and

Lutz. The root _ang_ is general Gmc., cp. O.E. _angmod_,

"vexed in mind." M.L.G. _anxt_, Germ. _angst_, Dan. _anger_.

The form of the word in Eng., however, is Scand.

ANGRYLY, _adv._ painfully. Wyntoun, VI, 7, 30. Deriv., cp. Cu.

angry, painful, O.N. _angrligr_, M.E. _angerliche_. The

O. Dan. vb. _angre_, meant "to pain," e.g., _thet angar mek,

at thu skal omod thorn stride_ (Kalkar).

APERT, _adj._ bold. Bruce, XX, 14. _apertly_, boldly, XIV, 77.

Evidently from O.N. _apr_, sharp, cp. _en aprasta hrið_,

Of Whales and Dwarves

"sharp fighting," cited in Cl. and V. Cl. and V. compares
  N.Ic. _napr_, "snappish," cp. furthermore _apirsmert_, adj.
  (Douglas, II, 37, 18), meaning "crabbed," the second element
  of which is probably Eng. _Apr_ in O.N. as applied to persons
  means "harsh, severe" (Haldorson).

ASSIL-TOOTH, _sb._ molar tooth. Douglas, I, 2, 12. See Wall.

AT, _conj._ that. O.N. _at_, Norse, Dan. _at_, to be regarded as a
  Scand. word. Might in some places be due to Celtic influence,
  but its early presence, and general distribution in Scand.
  settlements in England, Scotland, Shetland, etc., indicates
  that it is Scand.

BING, _sb._ a heap, a pile. Douglass, II, 216, 8. O.N. _bingr_,
  a heap, O. Sw. _binge_. Norse _bing_ more frequently a heap or

quantity of grain in an enclosed space. O. Dan. _byng_,
 bing.

BIR, BIRR, BEIR, _sb._ clamor, noise, also rush. S.S. 38; Lyndsay,
 538, 4280. O.N. _byrr_, a fair wind. O. Sw. _byr_. Cp. Cu.
 bur and Shetland "a pirr o' wind," a gust. Also pronounced
 bur, _bor_.

BIRRING, _pr. p._ flapping (of wings). Mansie Wauch, 159, 33. See
 bir.

BLA, BLAE (blç), _adj._ blue, livid. Douglas, III, 130, 30;
 Irving, 468. O.N. _blá_, blue, Norse _blaa, blau_, Sw. _blå_,
 Dan. _blaa_. Not from O.E. _blço_.

BLABBER, _vb._ to chatter, speak nonsense. Dunbar F., 112. O.N.
 blabbra, lisp, speak indistinctly, Dan. _blabbre_ id., Dan.
 dial. _blabre_, to talk of others more than is proper. M.E.

Of Whales and Dwarves

blaber, cp. Cu. _blab_, to tell a secret. American dial.

blab, to inform on one, to tattle. There is a Gael.

blabaran, sb. a stutterer, which is undoubtedly borrowed

from the O.N. The meaning indicates that.

BLAIK, _vb._ to cleanse, to polish. Johnnie Gibb, 9, 6. O.N.

blæikja, to bleach, O. Sw. _blekia_, Sw. dial. _bleika_. All

these are causative verbs like the Sco. The inchoative

corresponding to them is _blæikna_ in O.N., N.N., _blekna_ in

O. Sw., _blegne_ in Dan. See _blayknit_. Cp. Shetland _bleg_,

sb. a white spot.

BLAYKNIT, _pp._ bleached. Douglas, III, 78, 15. O.N. _blæikna_, to

become pale, O. Sw. _blekna_, Norse _blæikna_ id. O.N.

blæikr, pale. Cp. Cu. _blake_, pale, and _bleakken_ with

_i_-fracture. O.E. _blâc, bl¿can_.

BLECK, _vb._ put to shame. Johnnie Gibb, 59, 34, 256, 13. O.N.
 blekkja, to impose upon, _blekkiliga_, delusively,
 blekking, delusion, fraud; a little doubtful.

BLETHER, BLEDDER, _vb._ to chatter, prate. O.N. _blaðra_, to talk
 indistinctly, _blaðr_, sb. nonsense. Norse _bladra_, to
 stammer, to prate, Sw. dial. _bladdra_, Dan. dial. _bladre_,
 to bleet. Cp. Norse _bladdra_, to act foolishly.

BLATHER, _sb._ nonsense. Burns 32, 2, 4 and 4, 2, 4. O.N. _blaðr_,
 nonsense. Probably the Sco. word used substantively.

BLOME, _sb._ blossom. Bruce, V, 10; Dunbar, I, 12. Same as Eng.
 bloom from O.N. _blómi_.

BLOME, _vb._ to flourish, successfully resist. Douglas, IV, 58, 25.
 "No wound nor wapyn mycht hym anis effeir, forgane the speris

Of Whales and Dwarves

    so butuus blomyt he." Small translates "show himself
      boastfully." The word _blómi_ in O.N. used metaphorically
    means "prosperity, success."

BLOUT, BLOWT, _adj._ bare, naked, also forsaken. Douglas, III, 76,
    11; IV, 76, 6. O.N. _blautr_, Norse _blaut_, see Cl. and V.
    The corresponding vowel in O.E. is _ea_: _blçat_. The O.N.
    as well as the N.N. word means "soft." The O.E. word means
    "wretched." In Sco. _blout_ has coincided in meaning with
    blait. The Dan. word _blot_ is, on account of its form, out
    of the question.

This is a fraction of his findings, which are worth examining in their entirety. But they do confirm that Old Norse was the major source for the formation of Early Scots. The following extract shows his findings point to Norwegian sources not Danish as the origin for our Scandinavian words.

The general character of the Scand. loanwords in Sco. is Norse, not
Dan. This is shown by (a) A number of words that either do not exist
in Dan. or else have in Sco. a distinctively W. Scand. sense;
(b) Words with a W. Scand. form.

(a). The following words have in Sco. a W. Scand. meaning
or are not found in Danish:

AIRT, to urge. O.N. _erta_. Not a Dan. word.
APERT, boldly. O.N. _apr_. Not Dan.
AWEBAND, a rope for tying cattle. O.N. _háband_. Meaning
distinctively W. Scand.
BAUCH, awkward. Not E. Scand.
BEIN, liberal. Meaning is W. Scand.
BROD, to incite. O.N. _brodda_, id. Dan. _brodde_, means "to
equip with points."
BYSNING, monstrous. O.N. _bysna_. Not E. Scand.
CARPE, to converse. Not E. Scand.

Of Whales and Dwarves

CHOWK, jawbone. Rather W. Scand. than E. Scand.

CHYNGILL, gravel. A Norse word.

DAPILL, gray. A W. Scand. word.

DYRDUM, uproar. W. Scand. The word is also found in Gael.

Furthermore the form is more W. Scand. than Dan. Cp. _dýr_ and
dør.

DOWLESS, worthless. _Duglauss_ a W. Scand. word.

DUDS, clothes. Not found in Dan. or Sw.

ETTLE, aim at. W. Scand. meaning. O. Dan. _ætlæ_ meant "ponder
over."

FARRAND, handsome. This meaning is Icelandic and Norse.

FELL, mountain. W. Scand. more than E. Scand.

GANE, be suitable. O.N. _gegna_. Vb. not found in Dan.

GYLL, a ravine. O.N. _gil_. Is W. Scand.

HEID, brightness. O.N. _hærð_. Icel. and Norse.

HOOLIE, slow. O.N. _hógligr_. Not in Dan. or Sw.

KENDILL, to kindle. Ormulum _kinndlenn_ is from O. Ic. _kendill_
(Brate).

LIRK, to crease. I have not found the word in E. Scand.

MELDER, flour. O.N. _meldr_. Is W. Scand., particularly Norse.

POCKNET, a fishnet. O.N. _pôki-net. _ Not Dan.

RAMSTAM, indiscreet, boisterous. Both elements are W. Scand.

SCARTH, cormorant. W. Scand.

TARN, a lake. Distinctively Norse.

TYNE, to lose. O.N. _týna_. Distinctively Norse.

WAITH, booty. O.N. _væiðr_. Icel. and Søndmøre, Norway.

WARE, to spend. N. _verja_. W. Scand.

WICK, to cause to turn. O.N. _vikja_. Not Danish.

Bibliography

Old Norse Dictionary, Richard Cleasby and Gudbrand Vigfusson, 1874, Germanic Lexicon Project.

Anglo-Saxon Dictionary, Joseph Bosworth, and T.Northcote Toller, 1898 and 1921.

An Etymological Dictionary of the Gaelic Language, Alexander MacBain, 1911.

Concise Oxford Dictionary.

Welsh Meta-Dictionary, 'Yourdictionary.com'.

Concise Scots Dictionary.

Dictionary of the Scots Language.

Celtic Placenames of Scotland, W.J.Watson, 1993.

Place-Names of Scotland, 1934, James B. Johnston.

Scottish Place-Names, 1976, W.F.H. Nicolaisen.

Scotland's Place-names, 1995, David Dorward.

Statistical Accounts of Scotland, 1791-99, 1845.

Counties of Scotland, 1580-1928, National Library of Scotland.

Viking Place Names of East Lothian

Ragman Rolls

The Annals of Dunfermline

Bede's Ecclesiastical History

Anglo-Saxon Chronicle

Orkneyinga Sagas

Annals of Ulster

Annals of Tigernac

Pictish Chronicle,

Liber de Sancti Marie de Melrose

The Vikings in Ireland and Scotland in the ninth century, Professor Donnchadh O Corráinn, University College Cork.

Biography

Iain Johnstone was born in Edinburgh and studied at Edinburgh University. He taught in several schools in the Lothians, including East Lothian, where he now lives with his wife—and Rona.

Viking Place Names of East Lothian

Other books by the author

Acheson's Haven. *King James and the Witches.* A historical novel set in 16th.c. Scotland at the start of the infamous witch hunts in the Lothians, instigated by James V1. John Cunningham was a real schoolteacher at Prestonpans at this time and the novel follows the outline of his life using the surviving trial records.

Château de Carzac. The sequel to Acheson's Haven, this novel follows the fortunes of the friends of John Cunningham, who start a new life in the Medoc, in France. It is an exciting and dangerous time during the reign of Henry 1V of France. It also features the murderous Border warfare that existed in Scotland, culminating in the showdown between Johnstones and Maxwells at the Bloody Sands in Lockerbie.

Place Names of Scotland. *Fact and Fiction.* A general overview of Scots Gaelic, Brythonic and Norse place names, with an emphasis on the latter in the southern half of Scotland, where they are supposed to be rather scarce. This book challenges the accepted Scots history, which has resulted in many fictitious meanings given to place names. In the process, the Anglo-Saxon Heresy, mentioned in Robert Louis Stevenson's last letter to his brother in 1894, is exposed. It also has important consequences for the origin of the Scots non-Celtic tongue, which Scots Dictionaries must address. To be published soon.

All published by **Tarmagan Press**.

Viking Place Names of East Lothian

Of Whales and Dwarves

Postscript

Although this study has covered only a small part of East Lothian, my studies to date show the same pattern throughout East Lothian and the rest of the Lothians and Southern Scotland, which is detailed in my forthcoming book on Scots place names.

'Scots' place name societies, dictionaries and conventional histories will have to come to terms with my findings before there can ever be a Scots Dictionary of Place Names or an acceptable Scots Language Dictionary. It will be a mammoth task, but the sooner started the better.

Index

A

Abercorn 76
Aberlady 38
Achingall 44
Aðalsteinn 58
Aedan 78
Aedan macGabran 78
Æðelstan 58
Aesir 70
Aethelstan 17
Alba 13
Aldham 102
Aldhame 91
Aldhammar 105
Alfheim 74
Alfs 74
Amlaib 17
Anglecyn 41
Angus 16, 39
Anlaf 17
Annals of Lindisfarne 84
Asgard 70, 72
ASH 67
ASH (Anglo-Saxon Heresy) 25
Athelstaneford 58, 64, 109
Auda Ketilsdotir 75
Auldhame 106
Ayr 33

B

Balder 54, 70, 72
Baldr 70, 88
Baldred 28, 70, 91
Baldred of Kent 105
Baldred/Baldr 88
baldredi 28
Baldur 70
Balgone 100
Ballencrief 44
Balthere 92
Baltherus 91
Bamborough 16
Bangly Hill 58
Barn, Barni 63
Barnes 63
Barnes Castle 63
Barney Mains 63, 101
Bass Rock 87, 90, 96
Bassi 50, 96
Bebbanburch 16
Bede 35, 41, 91
Belhaven 28
Belton 28
Benedictine nunnery of St. Mary 97
Ber 50
Bernicia 15
Bernicia/Northumberland 40
Berryfell Hill 23
berserkers 50
Berserkr 96
Bersi 50

Of Whales and Dwarves

Beruvík 34
Berwick 34, 96
Berwick on Tweed 96
Biel 28
biel 28
Biel, bield, beel, bel 28
Biel Grange 28
Biel Hill 28
Biel Mill 28
Biel Water 28
Biorn 50
Birna 50
Bjarni 63
Bjorn 50
bœli 28
Blóts 104
blottoed 104
Bogside 33
Bonnington 100
Bœr 60
Briery Bank 80
Broadgate 26
Brockholes 86
Brok 73
Brók 86
Bromborough 17
Broxburn 86
Broxmouth 86
Brunanburh 17, 39
Brunnr 46
brunnr 26
Brunt Hill 46
Brythons 13, 14
Bucket Rocks 96
Bunkle Kirk 75

Burn 26
bùrn 46
bùrn 26
Burntisland 46
Butterdean Wood 83
Butterdenn 83
Byr 60
Byres Farm 60
Byres Hill 58, 60

C

Canongate 25
Car Rocks 89
Carlekemp 98
Carlops 24
Carmichael of
    Haddington 99
Carperstanes 98, 100
Catriona 89
ceann 14
Cluan 33
Coldale 47
Constantine 17
Cospatrick 39
Craigleith 95
Crann 14
Cruithne 13
Cunningham 33

D

Dalriada 13
darn 32
Darnaig, 32
Darncrook 33

Darnick 32
David I 18, 92, 104
David Dorward 25
Denglynhusse 65
Dictionary of the Scots Tongue 22
din 14
Din Eidyn 15
Din Guayroi 16
Dingle 65
Dingleton 32, 58, 65
Dinpaladyr 37
Dirleton 98
Domnal Breac 16
Dorothy Dunnett 18
Dublin 17
Dumbarton 33
dun 14
Dunbar 39, 44, 114
Duncan 11 92
Duncan 11, king of Scots 84
Dundee 13
Dunechtan 35
Dunholm 91
Dunpelder 114
Dupenderlaw 37

E

Eadbert 35, 39
Eadwine 16
Eanfrith 35
Earl Thorfinn 42
East Linton 48, 57
Ecgbert 35
Eddas 114

Edinburgh Castle 52
Edington 78
Edin's Hall 69
Ednam 78
Edwin 40, 84
Ekwall 44
Elvingston 74
Etain 16
Etan 16
Eyrr 33

F

Fagr 30
Fairlie 30
Falkirk 76
Fatt 44
Fattlipps 44
Fenrir 73
Fenton Barns 100
Fenwick 33
Fidra 98
Firth of Forth 15
Fishwick 33
fjall- hagar 23
Francis, Earl of Bothwell 47
Freki 54
Freya 70
Frigga 70
Fuylstrother 10

G

Gaels 14
Galashiels 28
Galawater 27
Gall/Gael 33

Galloway 33
Gamel 27
Gamelshiels 27
Garlabanck 48
Garleton 58, 98
Garleton monument 63
Garm 67, 72
Garmel 67
Garmeltun 67
Gata 25
Gate 25
Gegan 90
George Heriot 83
Geri 54
Gils 100
Gilsland 100
Gin Head 93
Ginglet 46
Glademoor 82
Gladsmuir 74, 82
Gleða 82
Glenmorisain 16
Gotterdammerung 73
Gourlay Bank 48
Gourlaybank 80
Govan. 76
Great Car 89
Gullane 98
Gullinbrusti 74
Gumishiels 27

H

Habbie's Howe 24
Hadda's stead 77
Hadden 77
Haddington 48, 57, 77
Hading's village 77
Hadingtoun 77
Hadyn 78
Hadynton 77
Haedentun 78
haga 23
Haga-land 23
Hagavík 23
Hailes 47
Hailes Castle 47
Hamar 102
Hamer 102
Hamer/Hamar 104
Hamfar 50
hamfar 114
Hammar 102
Hammer House 104
Hardgate 26
Haugh, hauch, haw, 23
Hawick 23
Heckie's Hole 87
Hedderwick 35, 48, 85
Heimskringla 52
Hel 67, 72
Henshiels 27
Herdman Flat 80
Heriot 83
Heugh 62, 96
Heugh, heuch 21
hlaew 10, 21
hœna 27
Hoð 71
Höfdingi 58
Höfdingitun 58

Högg 62
hogue 23
Hol, hola 24
Hoveden 92
Howe, how 24
Hugin 54
Hummel Ridges 98
Hummelknows 23
Hvin 101
Hvitingr 115
Hwita 111

I

Inga 85
Iona 35, 78
Isle of May 25, 85

J

Jagg 65
James VI 52, 99
Jocelin 38
John of Govan 78
Jól Blót 103
Jorvik 31
Jotunheim 69

K

Kaeheughs 58, 61
karl kempa 98
Kenneth MacAlpin 16, 39
Kentigern 90
Kepduff 38
Kerry 65
Kidnapped 28

Kilduff 38
Kilmodan 79
Knowe, know 25
kóg-bjarn 29
krókr 33
Kunningi 33

L

Ladies Skerrs 96
lægi 30
Lammermuirs 57
Landnamabok 23, 114
Law 10, 21, 98
Law Rocks 98
Lea, lie, ly 30
Leckerstanes 93
Leckmoram 94
Leckmoram Ness 93
Leith 94
Leith Roads 94
Leithies 94
Letham 83
Liber Sancti Marie of Melrose 91
Lif and Lifthrasr 73
Lindisfarne 78, 91
Linton 104
lippie 44
Lit 115
Loki 54, 67, 71, 72
Longskelly Pt 98
Lord of the Alfs 74
Luggate 26, 114
Luggate Burn 114
Luggate Road 114
lyggja 30

M

MacBeth 18, 42
Maes Howe 24
Malcolm Canmore 18
Mark-leið 99
Markle 48
Marly Knowe 99
Mary, queen of Scots, 47
Mead 55
Meadowbank 55
Meikle Spiker 85
Meiklerig 45
Meiklerig Wood 45
Melrose 32, 65, 92
Melrose Abbey 78
Metempsychosis 53
Mill Burn 97
Mimir 69
Minehowe 24
Moorfoot Hills 57
Mœr 94
Munin 54
Muirkirk 36
Mungo 38
Mylne 46

N

Nainn 115
Nainnnwarr 115
Nar 115
Narwar 115
Needless 64
Nið 65
Nið-lauss 65
Nið-reising 65
Ninewar 11, 115
Norsemen 16
North Berwick 84, 95
North Berwick Witches 99
North Berwick witches 52
Northumberland 16
Nungate 25
Nungate Road 97
Nunraw 104

O

Odin
    51, 52, 69, 89, 100, 103
Odin and Frigga 70
Olaf of Norway 66, 103
Olaf the White 39, 75
Olafr Hviti 114
Öngull 40
Orkney 21
Orkney Sagas 34, 96
Orkneys 13
Oswald 78
Othin 69
Over Hailes 47

P

Peffer Sands 88
pen 14
Pencraig 11, 49
Picts 13
Plato 54
Podlie Craig 93
podlock 93
Poldrate 80

Port of Leith 26
Prenn 14
Pressmennan Wood 44
Preston Kirk 91
Prestonhaugh 104
Prestonkirk 104
Prestonpans 104
Prestwick 33
Pretanni 13
Pridain/Pryden 13
Prof Nicolaison 31
Prof. Ó'Corráinn 41
Prof. Watson 37, 93
Pythagoras 54

R

Ragnarok 73
Raith Burn 33
Ralph of Haddington 78
Rammer Wood 46
raprain 44
Ravensheugh Sands 88
Ringhorn 71, 92
Robert Louis Stevenson 10, 28, 42, 46
Robert of Mythyngby 78
Roðgeir 89
Rodgers 89
Romans 15
Roodlands 81
Rugged Knowes 95

S

Sancti Baldridi 105

Sandy Hirst 86
Sassan 41
Sassanach 41
Sassenach 15
Saxons 15
Scart 85
Scart Rock 85
Scoughall 98
Scoughall Rocks 88
Sharper's Head 96
Shaw 29
Shawhill 33
Shiel, shiels, sheel 27
Sidegate 26
Siege of Dumbarton 38
Sigurd's Howe 24
Simeon of Durham 35, 84, 91
skaga 29
skáli 27
Skid Hill 56, 59
Skidbladnir 74
Skiði 61
skjól 27
skóg(r) 29
skóg-land 29
Skuggi 88
Skuggibarrin 88
Sleipner 54
Sleipnir 72
Slitrig Water 23
Smeaton 58, 65
Smið 59
Smyrton 66
Smyrtoun 65
Snorri Sturluson 52, 69

Snorri's Edda, Voluspo 115
Spittal Rigg 83
Spott Burn 86
Spott Wood 46
Spotti 46
St. Aidan 79
St. Aidan and St. Mary 79
St. Andrew's Kirk 99
St. Baldred's Boat 90
St. Baldred's Cradle 88
St. Baldrid's Well 105
St. Baldrid's Whill 105
St. Kentigern 38
St. Kilda 86
St. Mary 79
St. Mary's, Haddington 101
St. Mary's Whitekirk 101
St. Modans 79
St. Patrick 74
Steenstoun 46
Stefanson 46
Steinton 43
Stenton 43, 100
Stevenson House 46
Stevenson Mains 46
Strathclyde 17, 39
Sygn 73

T

Talorcan 35
tarf 33
Tarshaw 33
Tennyson 17

The Brunt 46
Thenaw 38
Thor 73, 103, 105
Thorfinn 18
Thorter Burn 104
Thurston 104
Timothy Pont 37, 44, 116
Ton, toun, toon, tun 29
Torness 104
Trabroun 44, 83
Tranent 44
Traprain Law 11, 37
tún 14
Tyinguham 91
Tyn 85
Tyne 57, 84
Tyninghame 84
Tynninghame 102
Tyr 73

U

Ugston 58
Úlfhamr 114
ut-hagi 23

V

Val-land 15
Valhalla 54, 70, 89
Valir 15
Valkyries 54
Vanir 70
Vár 116
Vár Blót 103
vátr 26

Vâtre 27
Ve 69
veittr 26
Viking Sagas 52
Vili 69
Volcae 15
Voluspo 73
Votadini 37

W

Water 26
Water of Leith 26
Watergate 26
Waughton 101
Wealas 15
wheen 36
Whin(g) Bauk 101
Whitberry Point 87
White 109
White Krist 103
Whitekirk 101
Whiting Bay 111
Whittingham 110
Whittingham(e) 11, 37
Whittinghame and
    Ninewar 108
Wick 31
William the Conqueror 18
William Wallace 15, 39
Wise Woman's Prophecy
    116
Wolfstar 9

Y

Ynglinga Saga 52
York 17, 31